Targeting of Drugs 3
The Challenge of Peptides and Proteins

NATO ASI Series

Advanced Science Institutes Series

A series presenting the results of activities sponsored by the NATO Science Committee, which aims at the dissemination of advanced scientific and technological knowledge, with a view to strengthening links between scientific communities.

The series is published by an international board of publishers in conjunction with the NATO Scientific Affairs Division

A	**Life Sciences**	Plenum Publishing Corporation
B	**Physics**	New York and London
C	**Mathematical and Physical Sciences**	Kluwer Academic Publishers
D	**Behavioral and Social Sciences**	Dordrecht, Boston, and London
E	**Applied Sciences**	
F	**Computer and Systems Sciences**	Springer-Verlag
G	**Ecological Sciences**	Berlin, Heidelberg, New York, London,
H	**Cell Biology**	Paris, Tokyo, Hong Kong, and Barcelona
I	**Global Environmental Change**	

Recent Volumes in this Series

Volume 232—Oncogene and Transgenics Correlates of Cancer Risk Assessments
edited by Constantine Zervos

Volume 233—T Lymphocytes: Structure, Functions, Choices
edited by Franco Celada and Benvenuto Pernis

Volume 234—Development of the Central Nervous System in Vertebrates
edited by S. C. Sharma and A. M. Goffinet

Volume 235—Advances in Cardiovascular Engineering
edited by Ned H. C. Hwang, Vincent T. Turitto,
and Michael R. T. Yen

Volume 236—Rhythms in Fishes
edited by M. A. Ali

Volume 237—The Photosynthetic Bacterial Reaction Center II: Structure,
Spectroscopy, and Dynamics
edited by Jacques Breton and André Verméglio

Volume 238—Targeting of Drugs 3: The Challenge of Peptides and Proteins
edited by Gregory Gregoriadis, Alexander T. Florence,
and George Poste

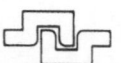

Series A: Life Sciences

Targeting of Drugs 3
The Challenge of Peptides and Proteins

Edited by

Gregory Gregoriadis and Alexander T. Florence

School of Pharmacy
University of London
London, United Kingdom

and

George Poste

SmithKline Beecham Pharmaceuticals
Surrey, United Kingdom

Springer Science+Business Media, LLC

Proceedings of a NATO Advanced Study Institute on
Targeting of Drugs: The Challenge of Peptides and Proteins,
held June 24 –July 5, 1991,
at Cape Sounion Beach, Greece

NATO-PCO-DATA BASE

The electronic index to the NATO ASI Series provides full bibliographical references (with keywords and/or abstracts) to more than 30,000 contributions from international scientists published in all sections of the NATO ASI Series. Access to the NATO-PCO-DATA BASE is possible in two ways:

—via online FILE 128 (NATO-PCO-DATA BASE) hosted by ESRIN, Via Galileo Galilei, I-00044 Frascati, Italy

—via CD-ROM "NATO-PCO-DATA BASE" with user-friendly retrieval software in English, French, and German (©WTV GmbH and DATAWARE Technologies, Inc. 1989)

The CD-ROM can be ordered through any member of the Board of Publishers or through NATO-PCO, Overijse, Belgium.

Library of Congress Cataloging in Publication Data

Targeting of drugs 3: the challenge of peptides and proteins / edited by Gregory Gregoriadis and Alexander T. Florence, and George Poste.
    p.   cm. – (NATO ASI series. Series A, Life sciences: v. 238)
  "Proceedings of a NATO Advanced Study Institute on Targeting of Drugs: the Challenge of Peptides and Proteins, held June 24–July 5, 1991 at Cape Sounion Beach, Greece" – T.p. verso
  "Published in cooperation with NATO Scientific Affairs Division."
  Includes bibliographical references and index.
  ISBN 978-1-4613-6276-0   ISBN 978-1-4615-2938-5 (eBook)
  DOI 10.1007/978-1-4615-2938-5
  1. Drug targeting – Congresses. 2. Protein drugs – Congresses. 3. Drugs – Vehicles – Congresses. I. Gregoriadis, Gregory. II. Florence, A. T. (Alexander Taylor) III. Poste, George. IV. North Atlantic Treaty Organization. Scientific Affairs Division. V. NATO Advanced Study Institute on Targeting of Drugs: the Challenge of Peptides and Proteins (1991; Ákra Soúnion, Greece) VI. Title: Targeting of drugs three. VII. Series.
  [DNLM: 1. Drug Carriers – congresses. 2. Drugs – administration & dosage – congresses. 3. Peptides – therapeutic use – congresses. 4. Proteins – therapeutic use – congresses. QV 38 T1855 1991]
  RM301.63.T38 1992
  615'.7 – dc20
  DNLM/DLC   92-48373
  for Library of Congress   CIP

ISBN 978-1-4613-6276-0

© 1992 Springer Science+Business Media New York
Originally published by Plenum Press in 1992

PREFACE

The NATO Advanced Studies Institute series "Targeting of Drugs" was
originated in 1981. It is now a major international forum, held every two
years in Cape Sounion, Greece, in which the present and the future of this
important area of research in drug delivery is discussed in great depth.
Five previous ASIs of the series dealt with drug carriers of natural and
synthetic origin, their interaction with the biological milieu, with ways by
which the latter influences such interaction and strategies by which milieu
interference curtailing the function of drug carriers is circumvented or
relieved. The present book contains the proceedings of the 6th NATO ASI
"Targeting of Drugs: The Challenge of Peptides and Proteins" held in Cape
Sounion during 24 June-5 July, 1991.

As the title implies, the book deals with a variety of approaches to
optimize the function of peptides and proteins used either as drugs and
vaccines or as drug delivery systems.

We express our appreciation to Ms. Brenda McCormack for her assistance
with the organization of the ASI and its co-ordination. We thank Dr. C.
Vakirtzi-Lemonias who, as chairperson of the Local Committee, contributed to
the success of the Institute. The ASI was held under the sponsorship of
NATO Scientific Affairs Division and co-sponsored and generously financed by
SmithKline Beecham Pharmaceuticals (King of Prussia). Financial assistance
was also provided by Syntex Laboratories Inc. (Palo Alto), Syntex Research
(Palo Alto) and Pfizer Central Research (Sandwich).

                                        Gregory Gregoriadis
                                        Alexander T. Florence
                                        George Poste

June 1992

CONTENTS

The Use of Derivatives of α-MSH for Targeting to Melanomas
    In Vivo   1
  D.R. Bard, C.G. Knight, D.P. Page-Thomas,
  E.P. Wraight and T.S. Maughan

The Development of Ricin A-Chain Immunotoxins for Clinical
    Trials in Patients with Hodgkin's Disease   9
  A. Engert and P. Thorpe

Ribosome-Inactivating Proteins from Saponaria Officinalis:
    Tools in the Design of Immunotoxins and Ligand Toxins   19
  M.R. Soria, L. Benatti, A. Ceriotti, A. Vitale and
  D.A. Lappi

Targeting with IgG and Immunoliposomes to Circulating Cells:
    The 'Target Cell Dragging' Concept   31
  D.J.A. Crommelin, P.A.M. Peeters and W.M.C. Eling

Liposome and Immunoliposome Mediated Delivery of Proteins
    and Peptides   45
  L. Huang and F. Zhou

Sterically Stabilized Liposomes as Drug Carriers:
    Pharmacokinetics, Tissue Distribution and Therapeutic
    Effects in Tumour-Bearing Mice   51
  A. Gabizon, S.K. Huang and D. Papahadjopoulos

Liposomes as Immunological Adjuvants   59
  G. Gregoriadis

Targeting Proteins to Antigen-Presenting Cells and Induction of
    Cytokines as a Basis for Adjuvant Activity   69
  A.C. Allison and N.E. Byars

Oral Administration of Peptides: Bypassing a Hostile Milieu   81
  M. Saffran

Oral Administration of Insulin: Imitating the Natural Pathway   89
  M. Saffran

Neuropeptide-Mediated Growth of Normal and Cancer Cells:
    Inhibition by Broad Spectrum Antagonists   97
  E. Rozengurt and T. Sethi

Bacterial Vectors to Target and/or Purify Polypeptides   109
  M. Hofnung, A. Charbit, J.-M. Clement, C. Leclerc
  P. Martineau, S. Muir, D. O'Callaghan, O. Popescu and
  S. Szmelcman

Participants' Photograph   121

Contributors   123

Index   125

THE USE OF DERIVATIVES OF α-MSH FOR TARGETING TO MELANOMAS IN VIVO

D.R. Bard[*], C.G. Knight[*], D.P. Page-Thomas[*],
E.P. Wraight[#] and T.S. Maughan[+]

[*]Strangeways Research Laboratory, Worts Causeway, Cambridge,
CB1 4RN, U.K. [#]Department of Nuclear Medicine,
Addenbrooke's Hospital, Hills Road, Cambridge, CB2 2QQ, U.K.
[+]Department of Clinical Oncology & Radiotherapeutics,
Addenbrooke's Hospital, Hills Road, Cambridge, CB2 2QQ, U.K.
[+]Present address: Velindre Hospital, Whitchurch, Cardiff,
CF4 7XL, U.K.

INTRODUCTION

Malignant melanoma in man is noted for early metastasis and
resistance to most conventional therapies. These properties make the
necessity for targeted agents particularly urgent. An accurate imaging
agent would assist the surgeon and the efficient targeting of cytotoxics
would enable their therapeutic ratios to be significantly improved.

Hitherto, most effort has been concentrated on the use of monoclonal
antibodies as targeting agents. High degrees of diagnostic accuracy have
been achieved with various antibodies, mostly directed against the 250 kD
proteoglycan, melanoma-associated antigen (MAA). For example, Salk and
coworkers (1988) were able to detect 82% of tumours over 3 cm in diameter
in a multicentre trial using an antibody (NR-ML-05) to MAA. Similar
results have been obtained using Fab fragments of other antibodies to the
same antigen (Eary et al, 1989; Lamki et al., 1990).

The routine clinical use of monoclonal antibody targeting is,
however, precluded by a number of difficulties. All the antibodies which
have so far reached clinical trial are of murine origin and their use is
accompanied by the generation of anti-murine antibodies in the majority
of patients (Lamki et al., 1990, Blottiere et al., 1990). Thus their
repeated use in the same patient would result in increasing levels of
immune elimination. A significant minority of patients also show toxic
effects (Dillman et al., 1988). Other problems include the stability and
reproducibility of labelling (Paik et al, 1985) and the slow rate of
extravasation of these large macromolecules and their restricted
penetration of tissue at the tumour site (Kwok et al., 1988; Ong & Mattes,
1989).

Both normal and malignant melanocytes possess receptors for the
tridecapeptide, α-melanocyte stimulating hormone (MSH) (Siegrist et al.,
1988, Ghanem et al., 1988; Tatro et al, 1990). The amino-acid sequence of
MSH is conserved in the seven mammalian species so far investigated and

Targeting of Drugs 3: The Challenge of Peptides and Proteins
Edited by G. Gregoriadis et al., Plenum Press, New York, 1992

1

this sequence is also found in some lower vertebrates such as the salmon (Eberle, 1988). α-MSH might therefore be expected to be invisible to the immune system of humans. Its low molecular weight (1665 D) should, furthermore, make it readily diffusible through tissues.

Native MSH has an acetyl group at the N-terminus. This may be replaced by much larger substituents such as biotin (Chaturvedi et al, 1984) or fluorescein (Chaturvedi et al., 1985) with little effect on hormonal activity. The N-terminal position was therefore chosen for the attachment of DTPA.

We have synthesised conjugates of MSH and the chelator, diethylenetriamine pentaacetic acid (DTPA) and assessed the efficacy of these compounds in targeting ^{111}In to melanomas (Bard et al., 1990a,b; Wraight et al., 1992).

During the last decade a number of so-called 'superpotent' analogues of MSH have been described. These have increased and prolonged activity compared with the native hormone and are more resistant to proteolysis (Sawyer et al., 1980, Wilkes et al, 1984 Al-Obeidi et al., 1989). Two analogues of one of these, Ac-Nle-Asp-His-DPhe-Arg-Trp-Lys-NH$_2$ (MSH (4-10)), in which the acetyl group was replaced with DTPA, were also prepared.

SYNTHESIS OF DTPA-MSH PEPTIDES

Peptides were synthesised step-wise from the C-terminus by the Fmoc-polyamide method (Atherton & Sheppard, 1989) on pepsyn K resin using a benzhydrylamine linker (Bard et al., 1990a). On cleavage of the peptide from the resin under acid conditions, the benzhydrylamine linker generates an amidated C-terminus (Bernatowicz et al., 1989).

After the peptide had been assembled on the resin, but before cleavage from the resin and deprotection, the N-terminal amino groups were reacted with DTPA bis-anhydride. DTPA bis-anhydride is a potential cross-linking reagent and the resultant product is dependent on the length of the peptide chain. If the native tridecapeptide is reacted with DTPA bis-anhydride, coupling between adjacent peptide chains occurs, giving bisMSH-DTPA (Fig. 1a). With the heptapeptide, MSH(4-10), however, monosubstitution of DTPA is favoured and monoMSH(4-10) (Fig. 1b) is the product. This effect appears to be determined by the length of the peptide chain and cannot be modified by altering the stoichiometry of the reaction with DTPA bis-anhydride.

Utilising this observation, bisMSH(4-10)-DTPA was prepared by interposing a tetraglycine spacer arm between the resin and the benzhydrylamine linker. This spacer lengthened the peptide sufficiently to enable crosslinking to take place. On cleavage, bisMSH(4-10)-DTPA (Fig. 1c) was released.

All peptides were purified by preparative reverse-phase chromatography. The final products were shown to be homogeneous by analytical HPLC and characterised by aminoacid analysis and fast-atom-bombardment mass spectrometry.

Peptides prepared by this method have small quantities of trifluoroacetate associated with them as a counteranion. Before the preparations were used in vivo this counteranion was replaced with acetate by reverse phase chromatography (Gabriel, 1987). The chelator

a   Ser-Tyr-Ser-Met-Glu-His-Phe-Arg-Trp-Gly-Lys-Pro-Val- NH$_2$
     /
DTPA
     \
      Ser-Tyr-Ser-Met-Glu-His-Phe-Arg-Trp-Gly-Lys-Pro-Val- NH$_2$

b

DTPA-Nle-Asp-His-DPhe-Arg-Trp-Lys- NH$_2$

c   Nle-Asp-His-DPhe-Arg-Trp-Lys- NH$_2$
     /
DTPA
     \
      Nle-Asp-His-DPhe-Arg-Trp-Lys- NH$_2$

Fig. 1. Structure of MSH chelator peptides:
a) BisMSH-DTPA
b) MonoMSH(4-10)-DTPA
c) BisMSH(4-10)-DTPA

peptides were shown to bind indium as previously described (Bard et al.,
1990).

ASSESSMENT OF HORMONAL ACTIVITY

The DTPA-peptides were assessed for hormonal activity by measuring
their ability to stimulate tyrosinase activity in cultures of the murine
melanotic melanoma cell line, Cloudman S91. Tyrosinase is the rate-
limiting enzyme in the biosynthesis of melanin and MSH stimulates a dose-
dependent increase in its activity in this cell line (Wong & Pawelek,
1973). The assay measures the formation of tritiated water from L-3,5-
^{3}H-tyrosine over a 24 h period (Lande et al, 1981, Bard et al., 1990a).
Results were expressed as EC50, the molar concentration of peptide
required to achieve half-maximum activity.

The EC50 values for bisMSH-DTPA monoMSH(4-10)-DTPA and their parent
peptides are given in Table 1. The activity of bisMSH-DTPA did not
differ significantly from that of native MSH. Mono-substitution of the
short-chain analogue had, however, a more marked effect and mono-MSH(4-
10)-DTPA showed a five-fold reduction in activity (P<0.001) when compared
with the acetylated parent peptide MSH(4-10)-Ac. Preliminary data
suggest that there is no significant difference in activity between
bisMSH(4-10)-DTPA and MSH(4-10)-Ac.

LOCALISATION IN VIVO

The Cloudman S91 cell line is syngeneic with the DBA2 mouse and
hence it was possible to measure the pharmacodynamics and biodistribution
of the MSH derivatives in mice bearing this tumour. Tumours were induced
either intradermally or intraperitoneally as previously described (Bard
et al., 1990a). The intradermal tumours were between 3 mm and 7 mm in

3

Table 1. Hormonal Activities of MSH Analogues

COMPOUND	*EC_{50}(nM)	
MSH	2.57	0.67
BisMSH-DTPA	3.18	0.42
MSH(4-10)-Ac	0.84	0.23
MonoMSH(4-10)-DTPA	4.40	0.74

*EC_{50} = Concentration required for half-maximum response.

Assays were carried out in using Cloudman S91 cells in monolayer culture. Results are the means of at least 4 independent determination ± s.e.m.

diameter (18-21 days growth) at the time of the experiments, whereas the intraperitoneal tumours were of varying size and distributed diffusely throughout the intraperitoneal cavity.

The chelator peptides, complexed with ^{111}In were injected intraperitoneally at a dose of 0.133 nmol peptide (0.123 MBq indium) per animal. Mono-MSH(4-10)-DTPA was cleared rather more rapidly from the circulation falling to 0.09 percent of the injected dose per gram of blood 4h after injection as compared with 1.32 for bisMSH DTPA.

All three compounds showed a greater uptake in the tumour than in spleen, skin, lung, blood, skeletal muscle, heart, eye or brain. The values for bisMSH-DTPA after 24 h are shown in Fig. 2. The only tissues which showed uptakes equal to or higher than tumour were liver and kidney. Tumour uptake of bisMSH-DTPA was halved when a 200 X molar excess of competing MSH was added at the same time as the labelled material, whilst uptakes in spleen, liver, kidney, skin, lung, heart, eye and brain were unaffected (Bard et al., 1990a). Preliminary data suggest that the corresponding uptake values for bis-MSH(4-10)-DTPA show a significantly higher uptake in tumour than in liver and lower kidney levels than those seen with bisMSH-DTPA.

Due to the more rapid clearance of mono-MSH(4-10)-DTPA tumour radioactivity became differentiated from the other tissues earlier than with the two bis compounds and a mean tumour/blood ratio of 10.6 ± 0.4 (n = 10) was measured after 4h (Bard et al., 1990b). With mono-MSH(4-10)-DTPA, the ratio of tumour to liver uptake was 3.58 ± 0.24 (n = 10) suggesting that, with this compound, liver metastases could be easily distinguished from their background in an imaging system.

CLINICAL TRIALS

The clinical application of bisMSH-DTPA as an imaging agent has been investigated in a trial involving fifteen patients at Addenbrooke's Hospital, Cambridge (Wraight et al., 1992). All the patients entered into the trial had either confirmed or suspected metastatic disease and gave informed consent to the procedures. Six of the fifteen patients were subsequently shown independently to be free of active disease at the time of the scan.

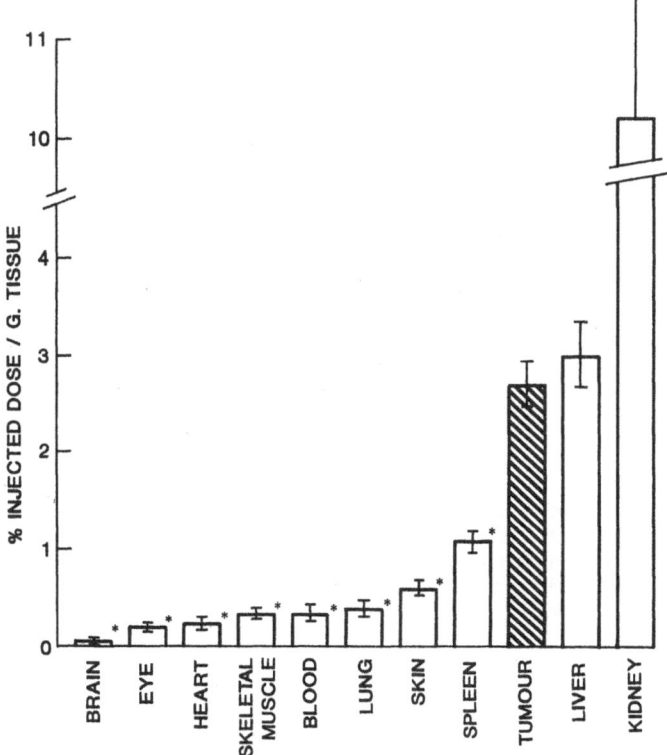

Fig. 2 Radioactivity in tissues of DBA2 mice 24 h after injection
of bisMSH-DTPA-[111]In. Results are expressed as percentages
of the original dose per gram of tissue ± s.e.m., n = 22.
Hatched bar = tumour, *Results significantly lower than
tumour (P<0.0001). Figure reproduced with permission of
British Journal of Cancer.

BisMSH-DTPA, 0.5mg (140 nmol), chelated with 40 MBq ^{111}In was
infused into each patient over a period of 30 min. Clearance of
radioactivity was measured with a collimated probe positioned at 3 m from
the patient and localisation of the isotope was visualised by means of
quantitative whole body gamma scans. In addition, spot scans were taken
of areas of interest. The scans were taken at intervals of 4 h, 24 h and
48 h after the completion of the injection. Blood samples were taken
immediately after injection and at 1 h, 2 h, 3 h, 24 h and 48 h. Due to
clinical constraints it was not always possible to take readings at each
time point on every patient.

In patients in which liver metastases were suspected, ^{99m}Tc colloid
subtraction imaging was performed at either 4 h or 24 h in addition to
the other procedures.

Radioactivity reached a maximum value of 16.9% of the injected
activity per litre within 15 min of the end of the infusion. It then
fell rapidly, reaching 2.84%/l at 4 h. Significant accumulation of
radioactivity was noted in the liver and kidney. The highest uptake was

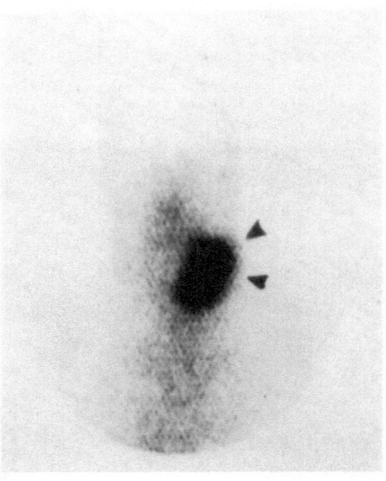

Fig. 3. Anterior gamma scan of the head and neck of a 54 year old
patient with a large submandibular melanoma metastasis 4 h
after infusion of 40MBq, 0.5mg ^{111}In-bisMSH-DTPA.
Extensive uptake is visible in the left submandibular area
(arrowed).

in the kidney which accounted for 25% of the initial whole body radioactivity at 24 h. The liver radioactivity at this time was 10% of the
initial whole body radioactivity (Wraight et al.,1992).

Known sites of disease could be clearly differentiated from the
surrounding tissues within 4 h of the completion of the infusion.
Although it was possible to continue imaging for at least 48 h no
significant improvement in resolution was obtained in the later scans.
Fig. 3 shows a tumour in the left submandibular area of a 54 year old man
who had had a primary tumour of the left eyelid removed four years
earlier. Deposits were identified in all of the nine patients with
active disease. In total, 41 lesions greater than 10 mm in diameter were
imaged out of a total of 46 (89%). The method was not able to detect
tumours smaller than this. Of the tumours detected, a total of eight had
not previously been suspected, but were subsequently confirmed by
ultrasound or computerised X-ray tomography. In two patients with
suspected liver metastases, bisMSH-DTPA was responsible for reversing
earlier positive diagnoses. Both these patients were still alive 30
months and 18 months respectively after the scans.

These data indicate that bisMSH-DTPA is an accurate imaging agent
for melanoma and a potentially useful tool for the localisation of
melanomas in human patients. Preliminary results suggest that the method
might be improved still further with the use of compounds based on short-
chain analogues of MSH.

Acknowledgements

This work was supported by the Cancer Research Campaign. The
attendance of DRB at the ASI was made possible by a travel grant from the
Cancer Research Campaign.

REFERENCES

Al-Obeidi, F., Hadley, M.E., Pettitt, B.M. and Hruby, V.J., 1989,
    Design of a new class of superpotent cyclic alpha-melanotropins
    based on quenched dynamic simulations, J. Am. Chem. Soc.,
    111: 3414.
Atherton, E. and Sheppard, R.C., 1989, Solid phase peptide synthesis, a
    practical approach, IRL Press, Oxford, U.K.
Bard, D.R., Knight, C.G. and Page-Thomas, D.P., 1990a, A chelating
    derivative of α-melanocyte stimulating hormone as a potential
    imaging agent for malignant melanoma, Br. J. Cancer, 62: 919.
Bard, D.R., Knight, C.G. and Page-Thomas, D.P., 1990b, Targeting of a
    chelating derivative of a short-chain analogue of alpha-melanocyte
    stimulating hormone to Cloudman S91 melanomas,
    Biochem. Soc. Trans., 18: 882.
Bernatowicz, M.S., Daniels, S.B. and Koster, H., 1989, A comparison of
    acid-labile linkage agents for the synthesis of peptide C-terminal
    amides, Tetrahedron Lett., 30: 4653.
Blottiere, H.M., Douillard, J.Y., Koprowski, H. and Steplewski, Z., 1990,
    Immunoglobulin class and immunoglobulin G subclass analysis of
    human anti-mouse antibody response during monoclonal antibody
    treatment of cancer patients, Cancer Res. (Suppl.), 50: 1051s.
Chaturvedi, D.N., Knittel, J.J., Hruby, V.J., de L. Castrucci, A.M. and
    Hadley, M.E., 1984, Highly potent peptide hormone analogues:
    synthesis and biological actions of biotin labelled melanotropins,
    J. Med. Chem., 27: 1406.
Chaturvedi, D.N., Hruby, V.J., de L. Castrucci, A.M. Kreutzfeld, K.L. and
    Hadley, M.E., 1985, Synthesis and biological evaluation of the
    superagonist [N'-chlorotriazinylamino-fluorescein-Ser^1, Nle^4,
    D-Phe^7]-α-MSH, J. Pharm. Sci., 74: 237.
Dillman, R.O., Beauregard, J.C., Jamieson, M., Amox, R.N. and Halpern,
    S.E., 1988, Toxicities associated with monoclonal antibody
    infusions in cancer patients, Mol. Biother., 1: 81.
Eary, J.F., Schroff, R.W., Abrams, P.G., Fritzberg, A.R., Morgan, A.C.,
    Kasina, S., Reno., J.M., Srinivasan, A., Woodhouse, C.S., Wilbur,
    D.S., Natale, R.B., Collins, C., Stehlin, J.S., Mitchell, M. and
    Nelp, W.B., 1989, Successful imaging of malignant melanoma with
    technetium-99m-labelled monoclonal antibodies, J. Nucl. Med.,
    30: 25.
Eberle, A., 1988, The Melanotropins, Karger, Basel.
Gabriel, T.F., 1987, A simple, rapid method for converting a peptide
    from one salt form to another, Int. J. Pept. Prot. Res., 30: 40.
Ghanem, G.E., Communale, G., Libert, A., Vercammen-Grandjean, A. and
    Lejeune, F.J., 1988, Evidence for α-melanocyte stimulating hormone
    (αMSH) receptors on human melanoma cells, Int. J. Cancer,
    41: 248.
Kwok, C.S., Cole, S.E. and Liao, S-K., 1988, Uptake kinetics of
    monoclonal antibodies by human malignant melanoma multicell
    spheroids, Cancer Res., 48: 1856.
Lamki, L.M., Zukiwski, A.A., Shanken, J., Legha, S.S., Benjamin, R.S.,
    Plager, C.E., Salk, D.F., Schroff, R.W. and Murray, J.L., 1990,
    Radioimaging of melanoma using ^{99m}Tc-labelled Fab fragment
    reactive with a high molecular weight antigen, Cancer Res.
    (suppl.), 50: 904s.
Lande, S., Pawelek, J., Lerner, A.B. and Emanuel, J.R., 1981, Assay of
    melanotropic peptides in an in vitro mammalian system,
    J. Invest. Dermatol., 77: 244.
Ong, G.L., Mattes and M.J., 1989, Penetration and binding of antibodies
    in experimental human solid tumours grown in mice, Cancer Res.,
    49: 4264.

Paik, C.H., Hong, J.J., Ebbert, M.A., Heald, S.C., Reba, R.C. and
        Eckelman, W.C., 1985, Relative reactivity of DTPA, immunoreactive
        antibody-DTPA conjugates and nonimmunoreactive antibody-DTPA
        conjugates toward indium-111, J. Nucl. Med., 26: 482.
Salk, D. and the multicentre study group, 1988, Technetium-labelled
        monoclonal antibodies for imaging metastatic melanoma: results of
        a multicentre clinical study, Seminars Oncol., 15:608.
Sawyer, T.K., Sanfilippo, P.J., Hruby, V.J., Engel, M.H., Heward, C.B.,
        Burnett, J.B. and Hadley, M.E., 1980, 4-Norleucine,
        7-D-Phenylalanine-alpha melanocyte stimulating hormone: a highly
        potent α-melanotropin with ultra long biological activity,
        Proc. Natl. Acad. Sci. USA, 77: 5754.
Siegrist, W., Oestreicher, M., Stutz, S., Girard, J. and Eberle, A.N.,
        1988, Radioreceptor assay for α-MSH using mouse B16 melanoma
        cells, J. Receptor Res., 8: 323.
Tatro, J.B., Atkins, M., Mier, J.W., Hardarson, S., Wolfe, H., Smith, T.,
        Entwistle, M.L. and Reichlin, S., 1990, Melanotropin receptors
        demonstrated in situ in human melanoma, J. Clin. Invest., 85: 1825
Wilkes, B.C., Sawyer, T.K., Hruby, V.J. and Hadley, M.E., 1984,
        Comparative biological activities of potent active site analogues
        of α-melanocyte stimulating hormone. Effect of tyrosine
        substitution at position 4, Int. J. Pept. Protein Res., 23: 632.
Wong, G. and Pawelek, J., 1973, Control of phenotypic expression of
        cultured melanoma cells by melanocyte stimulating hormone,
        Nature New Biol., 241: 213.
Wraight, E.P., Bard, D.R., Maughan, T.S., Knight, C.G. and Page-Thomas,
        D.P., 1992, The use of a chelating derivative of alpha-melanocyte
        stimulating hormone for the clinical imaging of malignant
        melanoma, Br. J. Radiol., 65: 112.

THE DEVELOPMENT OF RICIN A-CHAIN IMMUNOTOXINS FOR CLINICAL TRIALS IN

PATIENTS WITH HODGKIN'S DISEASE

Andreas Engert[1] and Philip Thorpe[2]

[1]Medizinische Universitatsklinik I der Universität zu Köln,
5000 Köln 41, Germany
[2]Cancer Immunobiology Center, The University of Texas,
Southwestern Medical Center, Dallas, Texas 75235-8576, USA

INTRODUCTION

Hodgkin's disease is an ideal target for immunotoxin therapy for several reasons. First, the Hodgkin/Reed-Sternberg (H-RS) cells, which are the putative malignant cells in this disease, have surface markers which are present only on a small minority of normal lymphoid cells. These markers are activation antigens recognized by monoclonal antibodies against the IL-2 receptor alpha-chain (CD25) (Agnarsson and Kadin, 1989), CD30 (Stein et al., 1985), and a 70kDa protein recognized by the unclustered IRac antibody (Hsu et al., 1987). Second, the majority of cells found within Hodgkin tumors are non-malignant reactive cells. Thus, the number of cells that needs to be killed is small. Third, Hodgkin tumors are usually well vascularized (Kaplan, 1980), suggesting that access of the immunotoxin to the target cells should be easier than it is, for example, in solid tumors. Fourth, Hodgkin's disease responds well to chemotherapy. It is therefore feasible to eradicate bulky disease by conventional therapy and then administer immunotoxins to kill residual tumor cells.

In this chapter, we summarize our previous data on the evaluation of monoclonal antibodies of various CD cluster determinants as potential ricin A-chain immunotoxins for the treatment of Hodgkin's disease. Several of these antibodies had little crossreactivity with normal human tissue and strongly stain H-RS cells in more than 90% of patients with Hodgkin's disease. Furthermore, immunotoxins constructed by linking CD25, CD30 and IRac antibodies via the sterically hindered SMPT linker to deglycosylated ricin A-chain were extremely effective at killing L540 Hodgkin cells in vitro and at destroying Hodgkin tumors in mice. From this survey, we selected the most powerful immunotoxin, RFT5γ1.dgA, for Phase I/II trials in patients with relapsed Hodgkin's disease.

POTENCY OF IMMUNOTOXINS IN VITRO

Thirty two monoclonal antibodies were evaluated for their ability to form ricin A-chain immunotoxins for treating Hodgkin's disease. The panel is shown in Table 1. It included 22 anti-CD25 MoAbs, 5 Moabs against the CD30 antigen and various others against CD15, CD36, CD71, MHC

Targeting of Drugs 3: The Challenge of Peptides and Proteins
Edited by G. Gregoriadis et al., Plenum Press, New York, 1992

9

Table 1. Monoclonal antibodies tested in an indirect assay for their potential as ricin A-chain immunotoxins against L540 Hodgkin cells

Cluster	Antibody	Subclass	Epitope	Author	IC50 (M)
CD 15	IG10	M	n.d.	Bernstein	$> 10^{-8}$
CD25	7G7/B6	G2a	B	Nelson	2×10^{-10}
	L54	G1	B	Maino	2×10^{-10}
	L61	G1	A	Maino	2×10^{-10}
	L62	G1	C	Maino	2×10^{-10}
	2A3	G1	A	Maino	3×10^{-9}
	143-13	G1	B	Vilella	1×10^{-9}
	143-24	M	C	Vilella	1×10^{-9}
	B-B10	G1	A	Wijdenes	4×10^{-11}
	B-F2	G1	A	Wijdenes	2×10^{-11}
	B-G8	G1	A	Wijdenes	2×10^{-9}
	B-E1	M	B	Wijdenes	2×10^{-9}
	B-G3	G1	B	Wijdenes	4×10^{-11}
	B-E10	G1	B	Wijdenes	1×10^{-10}
	M-A251	G1	B	Rieber	8×10^{-9}
	11H2.7	G2a	B	Mawas	$> 10^{-8}$
	LOTac1	G2b	A	Ravoet	$> 10^{-8}$
	anti-Tac	G1	A	Waldman	2×10^{-10}
	CD25-4E3	G2b	B	Knapp	3×10^{-10}
	CD25-8D8	G1	A	Knapp	2×10^{-10}
	CD25-3G10	G1	A	Knapp	3×10^{-10}
	CD25-9G8	G1	B	Knapp	3×10^{-10}
	RFT-5γ1	G1	A	Janossy	2×10^{-11}
	RFT5γ2a	G2a	A	Janossy	3×10^{-11}
CD 30	Ki-1	G3	B	Stein	$> 10^{-8}$
	HRS-1	G2a	A	Pfreundschuh	$> 10^{-8}$
	HRS-3	G1	A	Pfreundschuh	5×10^{-11}
	HRS-4	G1	A	Pfreundschuh	5×10^{-11}
	Ber-H2	G1	A	Stein	6×10^{-11}
CD 71	120-2A3	G1	n.d.	Vilella	2×10^{-10}
MHC II	TDR-31.1	G1	n.d.	Bodmer	2×10^{-10}
unclustered	IRac	G1	n.d.	Hsu	3×10^{-11}

class II, and one antibody that binds to an unclustered 70kDa antigen present on interdigitating reticulum cells as well as Hodgkin and Reed Sternberg cells (Hsu et al., 1987). The screening was performed by an indirect assay in which the target cells are incubated first with the test antibody and then with a Fab'-GAM-IgG.dgA immunotoxin. The immunotoxin binds to the test antibody and enters the target cell by the same or a similar route as would a primary immunotoxin. Thus, the assay accurately predicts the ability of any given antibody to function as a ricin A-chain immunotoxin (Till et al., 1988). The antibodies with the highest potency in the indirect assay were subsequently linked via the

Table 2. Cytotoxicity of immunotoxins on L540 Hodgkin cells

Specificity	Immunotoxin	$IC_{50} \pm$ sd (nMol)	
CD25	RFT5γ1.dgA	0.007	0.001
	RFT5γ2a.dgA	0.023	0.010
	B-B10.dgA	0.042	0.021
	B-F2.dgA	0.060	0.020
CD30	HRS-3.dgA	0.090	0.008
	HRS-4.dgA	0.100	0.040
	Ber-H2.dgA	0.200	0.050
70kDa	IRac.dgA	0.010	0.002
Control	Ricin	0.006	0.002
	Ricin A	> 100	
	OX7.dgA	> 1000	

bifunctional linker SMPT (4-sucinimidyloxycarbonyl-α-methyl(2-pyridyl-dithio)toluene) to deglycosylated ricin A-chain and highly purified on a Blue Sepharose CL-6B column. The methods for immunotoxin preparation and purification have been described in detail elsewhere (Thorpe et al., 1988). As predicted by the indirect assay, all the immunotoxins were highly toxic to L540 cells. Table 2 lists the eight most potent immunotoxins, all of which inhibited the protein synthesis of L540 Hodgkin cells by 50% at a concentration (IC_{50}) of 2×10^{-10}M or less in 3H-leucine uptake assays. The most potent immunotoxin, RFT5γ1.dgA, had an IC_{50} of 7×10^{-12}M which is identical to that of ricin itself under the same experimental conditions. The other three CD25 immunotoxins, RFT5γ2.dgA, B-B10.dgA, and B-F2.dgA were 3 times, 6 times and 9 times less potent than RFT5γ1.dgA, respectively. The CD30 immunotoxins were generally less effective than the CD25 immuno-toxins. HRS-3.dgA and HRS-4.dgA were almost 14 times less potent than RFT5γ1.dgA and Ber-H2.dgA was 28 times weaker. The IRac.dgA immunotoxin recognizing an unclustered 70kDa antigen was the second most potent immunotoxin with an IC_{50} value of 1×10^{-11}M. The cytotoxic effect of all the immunotoxins was specific since the native antibodies and OX7.dgA, an immunotoxin that does not bind to L540 cells, were not toxic at 10^{-6}M.

The major factor determining the potency of CD25 and CD30 ricin A-chain immunotoxins is the affinity of the monoclonal antibody for the target antigen (Figure 1) rather than the epitope recognized. This was demonstrated by Scatchard and FACS analyses (Engert et al., 1990; Engert et al., 1991). Different conclusions have been drawn from other studies: Shen et al. (1988) found that both antibody affinity and epitope location determine the potency of CD22 immunotoxins. By contrast, Press et al. (1988) concluded that epitope location critically influences the potency of CD3 immunotoxins.

STAINING OF NORMAL HUMAN AND HODGKIN'S DISEASE TISSUES

Excluding crossreactivity with vital human organs is important in the evaluation of immunotoxins for possible clinical use. The staining of the antibodies that formed the eight most potent immunotoxins were therefore checked for reactivity with frozen sections of 29 normal human tissues using an enhanced indirect immunoperoxidase technique (Janossy

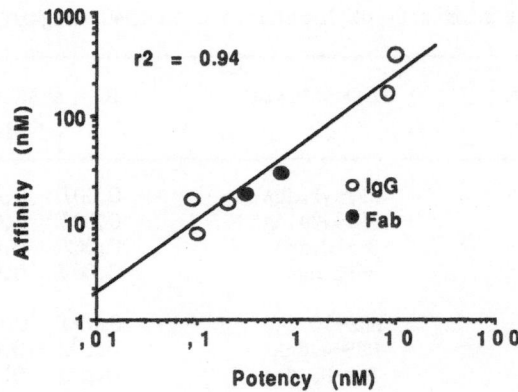

Fig. 1. Correlation of affinity of CD30 monoclonal antibodies and Fab' fragments with the potency of the corresponding ricin A-chain immunotoxin.

and Amlot, 1987). The results are summarized in Table 3. The CD25 monoclonal antibodies, RFT5γ1, RFT5γ2 and B-B10 as well as the CD30 antibodies HRS-3 and Ber-H2 had no major crossreactivity with any tissues other than lymphoid, where a few large cells in tonsils and lymph nodes were stained. When tested on sections of Hodgkin's disease tissue, CD25 and CD30 MAbs strongly stained nearly all cells that could be identified morphologically as H-RS cells in more than 90% of cases (data not shown).

The unclustered antibody IRac also had very little crossreactivity but preferentially stained nodular sclerosis and the mixed cellularity subtypes. In addition, the labelling was moderate to weak in most cases (data not shown). Therefore, an immunotoxin constructed with IRac might be of restricted use in clinical trials with Hodgkin's disease patients.

The CD25 antibody B-F2 which binds to the same epitope on the CD25 antigen as RFT5γ2 and B-B10 strongly stained the luminal surface of renal cells, columnar epithelial and mucus secreting cell of the uterine cervix. HRS-4 (CD30) recognizing the same epitope as HRS-3 and Ber-H2, unexpectedly stained normal pancreatic tissue. The strong cross-reactivity excludes B-F2 and HRS-4 from clinical use as immunotoxins. Evidently, antibodies recognizing the same epitope differ in their primary sequence in a way that can lead to spurious crossreactivity with other tissues. Similar findings have been observed with CD22 antibodies (Shen et al., 1988).

ANTITUMOR EXPERIMENTS IN MICE

Tumors were induced by subcutaneous injection of L540 Hodgkin cells into the right posterior gluteal region of so-called "triple beige" nude mice which have a B-cell deficiency in addition to the natural killer and T-cell defect known from beige nude mice. When the tumors had grown to approximately 0.7 cm in diameter, the mice were given a single intra-venous injection of immunotoxin. The dose of immunotoxin was 48 µg of total protein (of which 8 ug was ricin A-chain) corresponding to 40% of the LD_{50}. Figure 2a shows a typical experiment with the CD25 immunotoxins RFT5γ.dgA, RFT5γ2.dgA and B-B10.dgA. Figure 2b shows a similar experiment with IRac.dgA (unclustered), and the CD30

Table 3. Normal tissue staining patterns of antibodies to Hodgkin cell antigens

	CD25			B-F2	HRS-3	CD30		70KDa
	RFT5g1	RFT5g2a	B-B10			HRS-4	Ber-H2	IRac
Adrenal	-	-	-	-	-	n.d.	-	-
Brain (cortex)	-	-	-	-	-	n.d.	-	-
Brainstem	-	-	-	-	-	n.d.	-	-
Breast	-	-	-	-	-	.	-	-
Cerebellum	-	-	-	++	-	n.d.	-	-
Cervix	-	-	-	-	-*)	-*)	-*)	-
Colon	-	-	-	-	-	-*)	-	-
Gall bladder	-	-	-	-	-	n.d.	-	-
Heart	-	-	-	-	-	-	-	-
Ileum	-	-	-	++	-	-	-	-
Kidney	-	-	-	+++	-	-	-	-
Liver	-	-	-	-	-	-	-	-
Lung	-	-	-	-	-	-	-	-
Lymph node	-*)	-*)	-*)	-*)	-*)	-*)	-*)	-
Mucosa (nasal)	-	-	-	-	-	n.d.	-	-
Oesophagus	-	-	-	-	-	n.d.	-	-
Ovary	-	-	-	-	-	+++	-	-
Pancreas	-	-	-	-	-	n.d.	-	-
Parathyroid	-	-	-	-	-	-	-	-
Spleen	-	+	-	-	-	n.d.	-	-
Stomach (antrum)	-	+	-	-	-	-	-	-
Stomach (body)	-	-	-	-	-	n.d.	-	-
Testis	-	-	-	-	-	-	-	-
Thyroid	-	-	-	-	-*)	n.d.	-*)	-
Thyroid (AI)	-	-	-	-	-	n.d.	-	-
Thyroid (Hashimoto's)	-*)	-*)	-*)	-	-*)	-*)	-*)	-
Tonsils	-*)	-	-	-	-	n.d.	-	-
Uterus	-	-	-	-	-	n.d.	-	-
Vagina	-	-	-	-	-	n.d.	-	-

*) Rare cells within lymphoid tissue stain positively

13

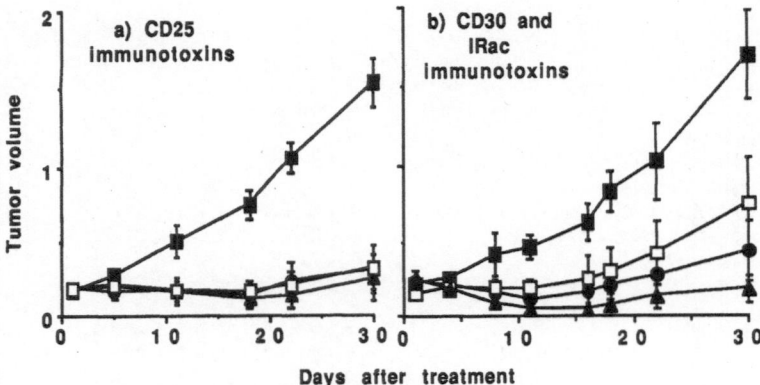

Fig. 2. Antitumor effects of IgG immunotoxins on solid L540
tumors in triple beige mice. Groups of 8-10 animals
received i.v. injections of the following: in a, PBS (■);
RFT5γ1.dgA (▲); RFT5γ2a.dgA (△); BB10.dgA (●); in b, PBS
(■); BerH2.dgA (□); HRS3.dgA (●); IRac.dgA (▲).

immunotoxins, HRS-3.dgA and Ber-H2.dgA. In Table 4 are summarized the
antitumor effects of these six immunotoxins as measured by the "growth
index" which is the mean tumor volume per group on day 30 after treatment
divided by the mean tumor volume on the day of treatment (day 1).
RFT5γ1.dgA was the most powerful immunotoxin (growth index 0.5) followed
by RFT5γ2.gA and IRac.dgA (growth index 0.8), HRS-3.dgA (growth index
1.4) and, lastly, Ber-H2.dgA (growth index 4.8). There were a
substantial number of complete remissions ranging from nearly 80%
(RFT5g1.dgA) to 27% (Ber-H2.dgA). Unconjugated antibodies, mixtures of
antibodies or A-chain, and immunotoxins of irrelevant specificity all
lacked significant antitumor activity, showing that the marked antitumor
effects of the anti-H-RS immunotoxins were due to immunotoxin-mediated
tumor cell killing.

These marked antitumor effects obtained with a single i.v. injection
of several different immunotoxins in the L540 Hodgkin model (Engert et
al., 1991, 1990) contrasts with the relative lack of efficacy of
immunotoxins in other solid tumor models. There have been exceptions
(Bernhard et al., 1983) but, generally, multiple injections at dosages
exceeding the LD50 (Byers et al., 1987) or intratumoral injection (Hara
et al, 1988) have been necessary. From localization experiments with
HRS-3 (data not shown) we have no evidence that Hodgkin's tumors in mice
are unusually permeable. Probably, the excellent antitumor activity
reflects the fact that we used highly purified, SMPT linked,
deglycosylated ricin A-chain immunotoxins. These improvements have been
shown previously to enhance the antitumor activity of immunotoxins
substantially in mouse T-cell (6) and B-cell (Fulton et al., 1988)
lymphoma models.

MUTANTS

Tumors from mice which had complete remissions after treatment with B-
B10.dgA (CD25), HRS-3.dgA (CD30), or IRac.dgA but which subsequently
relapsed were re-established in tissue culture and retreated. All four

Table 4. Antitumor effects of immunotoxins directed against HD-associated antigens in solid L540 tumor bearing mice

Cluster	Immunotoxin	Tumor Growth Index[*] (S.D.)	
CD25	RFT5g1.dgA	0.5	0.3
	RFT5g2.dgA	0.8	1.2
	B-B10.dgA	0.9	0.2
CD30	HRS-3.dgA	1.4	0.5
	Ber-H2.dgA	4.7	1.8
70kDA (unclustered)	IRac.dgA	0.8	0.2
	Untreated[**]	9.7	1.6

[*]   Ratio of tumor volume 30 days after treatment to volume on day of treatment

[**]   All other control groups (diluent, antibody alone, irrelevant immunotoxin) had tumor growth indices of 7.5 to 10.3

sublines that originated from relapsed IRac.dgA tumors were 40, 60 and 200 times less sensitive to IRac.dgA than the original L540 line (Table 5). The degree of resistance of the sublines to IRac.dgA correlated with the decrease in their ability to bind IRac as measured by FACS analyses. These results indicate that the solid L540 tumors contained a few IRac deficient mutants which were not killed by IRac.dgA and which caused a relapse at the original tumor site. Importantly, the IRac.dgA resistant sublines were as sensitive as the original L540 line for HRS-3.dgA or B-B10.dgA, suggesting that treatment of mice with a cocktail of two or more immunotoxins directed against different antigens might reduce the likelihood of mutant tumor cell escape.

CONCLUSIONS

Immunotoxins prepared by linking ricin A-chain to several MoAbs which recognize different markers on H-RS cells have potent antitumor activity in vitro and against solid Hodgkin's tumors in mice. Most CD25 and CD30 immunotoxins strongly stain H-RS cells in more than 90% of patients with Hodgkin's disease and have little crossreactivity with normal human tissues. The most powerful immunotoxin, RFT5γ1.dgA (CD25), is currently being scaled up for clinical trials in patients with Hodgkin's disease.

Table 5. Characteristics of L 540 sublines derived from IRac.dgA-
treated mice which had complete remissions but
subsequently relapsed

Sub-line	Resistant or Sensitive	Antigen density (% MFI)	Immunotoxin sensitivity IC_{50} (nM)		
			IRac.dgA	HRS-3.dgA	B-B10.dgA
1	Sensitive	87	0.02	0.10	0.03
2	Resistant	33	0.40	0.06	0.06
3	Resistant	24	0.60	0.10	0.04
4	Resistant	13	2.00	0.40	0.05
L540	Sensitive	100	0.01	0.10	0.05

References

Agnarsson, B. A. and Kadin, M. E., 1989, The immunophenotype of Reed-Sternberg cells. A study of 50 cases of Hodgkin's disease using fixed frozen tissue, Cancer, 63:2083.

Bernhard, M. I., Foon, K. A., Oeltmann, T. N., Key, M. E., Hwang, K. M., Clarke, G. C., Christensen, W. L., Hoyer, L. C., Hanna, M. G. and Oldham, R. K., 1983, Guinea pig line 10 hepatocarcinoma model: characterization of monoclonal antibody and in vivo effect of unconjugated antibody and antibody conjugated to diphteria toxin A-chain, Cancer Res., 43:4420.

Byers, V. S., Pimm, M. V., Scannon, P. J., Pawluzyk, S. I. and Baldwin, R. W., 1987, Inhibition of growth of human tumor xenographs in athymic mice treated with ricin toxin A-chain-monoclonal-antibody 791T/36 conjugates, Cancer Res., 47:5042.

Engert, A., Burrows, F., Jung, W., Tazzari, P. L., Stein, H., Pfreundschuh, M., Diehl V. and Thorpe, P., 1990, Evaluation of ricin A-chain containing immunotoxins directed against CD30 as potential reagents for the treatment of Hodgkin's disease, Cancer Res., 50:84.

Engert, A., Martin, G., Amlot, P., Wijdenes, J., Diehl V. and Thorpe, P., 1991, Immunotoxins constructed with CD25 monoclonal antibodies and deglycosylated ricin A-chain have potent antitumor effects against human Hodgkin cells in vitro and solid Hodgkin tumors in mice, Int. J. Cancer, 49:450.

Engert, A., Martin, G., Pfreundschuh, M. Amlot, P. Hsu, S. M., Diehl V. and Thorpe P., 1990, Antitumor effects of ricin A-chain immunotoxins from intact antibodies and Fab' fragments on solid human Hodgkin's disease tumors in mice, Cancer Res., 50:2929.

Fulton, R. J., Uhr, J. W. and Vitetta, E. S., 1988, In vivo therapy of the BCL tumor: effect of immunotoxin valency and deglycosylation of the ricin A-chain, Cancer Res., 48:2626.

Hara, H. Luo, Y. Haruta, Y. and Seon, B. K., 1988, Efficient transplantation of human T-leukemia cells into nude mice and induction of complete regression of the transplanted distinct tumors by ricin A-chain conjugates of monoclonal antibodies SNS and SN6, Cancer Res., 48:4673.

Hara, H., Luo, Y., Press, O.W., Martin, P. J., Thorpe, P. E. and Vitetta, E. S., 1988, Ricin A-chain containing immunotoxins directed against different epitopes on the CD3 molecule differ in their ability to kill normal and malignant T cells, J. Immunol., 141:4410.

Hsu, S.-M., Ho. Y.-S. and Hsu, P.-L., 1987, Effect of monoclonal anti- bodies anti-2H9, anti-IRac, and anti-HeFi-1 on the surface antigens of Reed-Sternberg cells, J. Nat. Cancer Inst., 79:1091.

Janossy, G. and Amlot, P., 1987, Immunofluorescence and immunohistology, in: C. G. B. Klaus, ed., "Lymphocytes: A Practical Approach", IRL Press, Oxford.

Kaplan, H.S., 1980, Hodgkin's disease: unfolding concepts concerning its nature, management and prognosis, Cancer, 45:2439.

Shen, G. L., Li, J. L., Ghetie, M. A., Ghetie, V., May, R. D., Till, M., Brown, A. N. F., Relf, M. G., Knowles, P., Uhr, J. W., Janossy, G., Amlot, P., Vitetta, E. E. and Thorpe, P. E., 1988, Evaluation of four CD22 antibodies as ricin-A-chain-containing immunotoxins for the in vivo therapy of human B-cell leukemias and lymphomas, Int. J. Cancer, 42:792.

Stein, H., Masan, D. Y., Gerdes, J., O'Connor, N., Wainscoat, J., Pallesen, G., Gatter, K., Falini, B., Delsol, G., Lembke, H., Schwarting, R. and Lennert, K., 1985, The expression of the Hodgkin's disease- associated antigen Ki-1 in reactive and neoplastic lymphoid tissue: evidence that Sternberg-Reed cells and histiocytic malignancies are derived from activated lymphoid cells, Blood, 66:848.

Thorpe, P. E., Wallace, P. M., Knowles, P. P., Relf, M. G., Brown, A. N. F. Watson, G. J., Blakey, D. C. and Newell, D. R., 1988, Improved anti-tumor effects of immunotoxins prepared with deglycosylated ricin A-chain and hindered disulfide linkages, Cancer Res., 48:6396.

Till, M., May, R. D., Uhr, I. W., Thorpe, P. E. and Vitetta, E. S., 1988, An assay that predicts the ability of monoclonal antibodies to form potent ricin A-chain-containing immunotoxins, Cancer Res., 48:1119.

RIBOSOME-INACTIVATING PROTEINS FROM SAPONARIA OFFICINALIS: TOOLS IN THE

DESIGN OF IMMUNOTOXINS AND LIGAND TOXINS

Marco R. Soria, Luca Benatti[*], Aldo Ceriotti[#],
A. Vitale[#], Douglas A. Lappi[+]

Department of Biotechnology, San Raffaele Research
Institute, Via Olgettina 60, 20133 Milano, Italy
[*]Biotechnological Research, Farmitalia Carlo Erba,
24 Via E. Bezzi, 20146 Milano, Italy
[#]Istituto Biosintesi Vegetali del CNR, Via Bassini 15
20133 Milano, Italy
[+]Department of Cellular Growth Biology, The Whittier Institute
for Diabetes and Endocrinology, La Jolla, CA, USA

INTRODUCTION

Single chain ribosome-inactivating proteins (RIPs) are denominated Type 1 RIPs as opposed to RIPs consisting of two nonidentical subunits (A and B chains) that are joined by a disulfide bond (Stirpe and Barbieri, 1986; Stirpe et al., 1992). Complete amino acid sequences are now known for many Type 1 and Type 2 RIPs. Several type 1 RIPs were isolated from Saponaria officinalis (Stirpe et al., 1983). Among these, saporin-6 (also called SO-6) has amino acid sequence similarity with the RIPs from Phytolacca americana and Phytolacca dodecandra (Lappi et al., 1985; Kung et al., 1990), and with dianthin-30, from Dianthus caryophyllus, which belongs to the Caryophyllaceae family like Saponaria officinalis (Legname et al., 1991).

Studies on the genetic organization and sequence of some type 1 RIPs and their transcripts are now slowly emerging. In our laboratory, we characterized several aspects of the genetic system coding for saporin-type RIPs of Saponaria officinalis.

CLONING STUDIES

A cDNA coding for saporin-6 was recently cloned in our laboratory (Benatti et al., 1989). The translated sequence showed the signal peptide and the coding region for saporin-6. Comparison of the amino acid sequence determined directly on saporin-6 purified from seeds (Maras et al., 1990) to the predicted amino acid sequence from the leaf cDNA showed complete identity between these sequences at all but few amino acid residues along the molecule (Fig. 1 b and d). The differences between amino acid residues predicted by cDNA cloning and those determined by direct sequencing might be due to different forms of saporin-6 in the same plants, as in the case of pokeweed antiviral protein, PAP (Houston et al., 1983), of ricin (Lamb et al., 1985), and of trichosanthin (Chow et al, 1990). The existence of several genes coding for various forms of saporin-6 is indicated by genomic

Targeting of Drugs 3: The Challenge of Peptides and Proteins
Edited by G. Gregoriadis et al., Plenum Press, New York, 1992

19

```
ATAYTLNLAN PSASQYSSFL DQIRNNVRDT SLIYGGTDVA VIGAPSTTDKF LRLNFQGPRG (a)
VTSITLDLVN PTAGQYSSFV DKIRNNVKDP NLKYGGTDIA VIGPPS-KEKF LRINFQSSRG (b)
VTSITLDLVN PTAGQYSSFV DKIRNNVKDP NLKYGGTDIA VIGPPS-KDKF LRINFQRTRG (c)
VTSITLDLVN PTAGQYSSFV DKIRNNVKDP NLKYGGTDIA VIGPPS-KDKF LRINFQRTRG (d)
                                                    (E)
                                                     *

TVSLGLRREN LYVVAYLAMD NANVNRAYYF KNQITSAELT ALFPEVVVAN QKQLEYGEDY (a)
TVSLGLKRDN LYVVAYLAMD NTNVNRAYYF RSEITSAEST ALFPEATTAN QKALEYTEDY (b)
TVSLGLKRDN LYVVAYLAMD NTNVNRAYYF KSEITSAELT ALFPEATTAN QKALEYTEDY (c)
TVSLGLKRDN LYVVAYLAMD NTNVNRAYYF KSEITSAELT ALFPEATTAN QKALEYTEDY (d)
                          (R)       *
                           *

QAIEKNAKIT TGDQSRKELG LGINLLITMI DGVNKKVRVV KDEARFLLIA IQMTAEAARF (a)
QSIEKNAQIT QGDQSRKELG LGIDLLSTSM EAVNKKARVV KDEARFLLIA IQMTAEAARF (b)
QSIEKNAQIT QGDKSRKELG LGIDLLLTFM EAVNKKARVV KNEARFLLIA IQMTAEVARF (c)
QSIEKNAQIT QGDKSRKELG LGIDLLLTFM EAVNKKARVV KNEARFLLIA IQMTAEVARF (d)
             *            * *            *               *

RYIQNLVTKN FPNKFDSENK VIQFQVSWSK ISTAIFGDCK NGVFNKDYDF GFGKVRQAKD (a)
RYIQNLVIKN FPNKFNSENK VIQFEVNWKK ISTAIYGDAK NGVFNKDYDF GFGKVRQVKD (b)
RYIQNLVTKN FPNKFDSDNK VIQFEVSWRK ISTAIYGDAK NGVFNKDYDF GFGKVRQVKD (c)
RYIQNLVTKN FPNKFDSDNK VIQFEVSWRK ISTAIYGDAK NGVFNKDYDF GFGKVRQVKD (d)
             * *          * *


LQMGLLKYLG RPK (a)
LQMGLLMYLG KPK (b)
LQMGLLMYLG KPK (c)
LQMGLLMYLG KPK (d)
```

Fig. 1.   Comparison of the deduced amino acid sequences
          (corresponding to the mature protein) of
          (a) dianthin-30 cDNA, (b) saporin-6 leaf cDNA
          (c) sap-2 genomic clone. The amino acid sequence
          reported by Maras et al (1990) is depicted in (d).
          Residues that differ between (a) and (b) are underlined.
          Asterisks mark residues that differ between (d) (c)
          and (b).

amplification experiments on leaf DNA using the polymerase chain reaction
(Soria, 1990).

THE 3' END OF SAPORIN-6 cDNA: A TARGETING-CODING REGION?

        Initially, we could not identify a translation termination codon at the
3' end of our cDNA clones coding for saporin-6, because they ended with a
"natural" EcoRI site at their 3' end, that is, a site not resulting from the
addition of the EcoRI linkers to the cDNA. In addition, positive
identification of the COOH-terminal end of saporin-6 was hindered by the
resistance of this protein to treatment with some proteases, including
carboxypeptidases (Stirpe et al., 1983; Soria et al., 1992).

        We have recently determined that the missing portion of the cDNA for
saporin-6 codes for a 22 amino acid carboxyl-terminal extension that is not
found in the mature protein, followed by the stop codon and polyadenylation
signal. This is evidenced by the positive identification of the COOH-
terminal residues as ...Pro-Lys by hydrazinolysis and by treating saporin-6
with carboxypeptidase P after thermal denaturation (Benatti et al., 1991).
Thus, saporin-6 derives by a processing mechanism from a longer precursor
whose cDNA extends beyond the 3' end of our cDNA clones. Therefore, all the
coding portion for the COOH-terminus of mature saporin-6 is contained in our
cDNA clones.

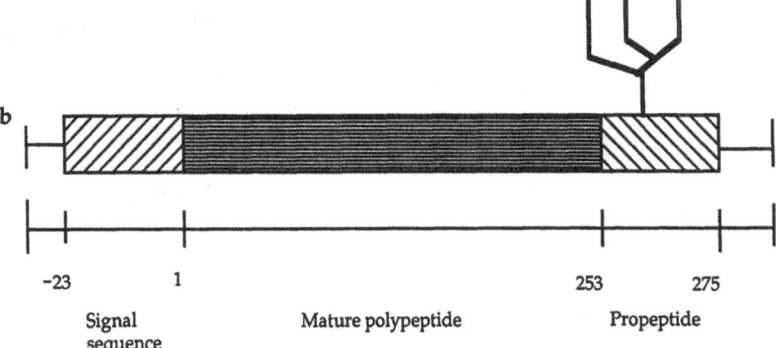

a
                ↓254   259   275
        K   P   K   S   S   N   E   A   N   S   T   V   R   H   Y   G   P   L   K   P   T   L   L   I   T stop
        AAACCAAAGTCGTCAAACGAGGCGAATTCTACCGTACGCCACTACGGTCCTCTGAAGCCTACTTTACTAATCACATGA

b

        -23   1   253   275

        Signal   Mature polypeptide   Propeptide
        sequence

Fig. 2.   The 3' end of saporin-6 cDNA encodes a carboxyl-
          terminal extension with an N-linked glycosylation
          site. (a) The 22-amino acid COOH-terminal extension
          of saporin and the corresponding nucleotide sequence.
          The arrow indicates the end of mature saporin-6. The
          potentially glycosylated asparagine residue (Asn 259)
          is underlined. (b) Structure of saporin cDNA. The
          putative N-linked glycosylation site (Asn 259-Ser-Thr
          261) is depicted by attachment of a glycan tree to
          the COOH-terminal extension.

A potential N-linked glycosylation site (Asn-X-Ser/Thr) was found to
reside within the propeptide located at the carboxyl-terminal end of saporin
(Fig. 2). The actual glycosylation of this site still needs to be demonstrated: not all potential glycosylation sites are glycosylated in vivo,
because often the conformation of the growing polypeptide hinders access to
oligosaccharide transferase (Kornfeld and Kornfeld, 1985). Pulse-chase
experiments performed on developing seeds of Saponaria officinalis indicate
that proteolytic processing events take place in seed saporin-6. The
protein was immunoprecipitated from pulse-labelled seeds, yielding an SDS-
PAGE pattern consisting of at least four polypeptides in the 30-35 KD range,
whereas seeds that had been labelled and then chased in cold medium to allow
processing of the newly synthesized proteins did not yield the slower-
migrating bands. A corresponding increase in intensity of the faster
migrating polypeptides was instead observed, confirming the existence of a
precursor-product relationship (Soria et al., 1992). Since mature saporin-6
is not glycosylated, it is likely that this change reflected the proteolytic
processing of the polypeptide chain with removal of a possibly glycosylated
propeptide and a reduction in MW of about 4000 daltons. However, we have
used leaf mRNA to produce and characterize our cDNA clones at the 3' end,
whereas saporin-6 protein is obtained and characterized from seeds. Thus,
it could be that the primary translation products of the leaf and of the
seed proteins differ in possessing or not the same carboxyl-terminal
extension as that described in Fig. 2. Indeed, the amino acid sequence
determined on saporin-6 purified from seeds (Maras et al., 1990) is almost
identical to the deduced amino acid sequence of sap-2, a saporin genomic
clone (Fordham-Skelton et al., 1991; Table 1 and Fig. 1).

Table 1. Comparison of the COOH-terminal propeptides of saporin-6, sap-2, dianthin, tobacco glucanase, barley lectin, chitinase, trichosanthin, momordin and tomato polygalacturonase (PG)

Propeptide	Protein	References
SSNEANSTVRHYGPLKPTLLIT	Saporin-6	Benatti et al., 1991
SSNEANSTAYATTVL	Sap-2	Fordham-Skelton et al., 1991
SSSIEANSTDDTADVL	Dianthin	Legname et al., 1991
VSGGVWDSSVETNATASLVSEM	Tobacco glucanase	Shinshi et al., 1988
LLVDTM	Chitinase	Payne et al., 1990
VFAEAIAANST--LVAE	Barley lectin	Bednarek et al., 1990
VFAEAITANST--LLQE G AT A	Wheat germ agglutinin	Bednarek et al., 1990
AMDDDVPMTQSFGCGSYAI	Trichosanthin	Chow et al., 1990
EGDNGDVSTTHGFSSY	Momorcharin	Ho et al., 1991
LEISEDEALLYNY	Tomato PG	Osteryoung et al., 1990

The N-linked glycosylation signal is underlined and conserved positions among RIPs and other sequences are marked with dots. Sequences are aligned, with dashes introduced, to maximize homology.

In Table 1 we have compared the carboxyl-terminal extension of saporin-6 to those of sap-2 and of dianthin-30, deduced from a leaf cDNA (Legname et al., 1991). The carboxyl-terminal propeptides of tobacco glucanase and chitinase, of wheat germ agglutinin and of barley lectin are also compared. The propeptides of saporin-6, Sap-2 and dianthin-30 have evident homology at their NH$_2$-termini, down to the common glycosylation site at position 5 or 6. After this, there is a marked divergence in primary sequence as well as in charge. The presence of an N-linked glycosylation site is also common to the propeptides of glucanase, wheat germ agglutinin and barley lectin. In these three proteins the site is glycosylated in vivo. In addition, in glucanase the region ahead of the glycosylation site has homology to the same region in the three RIPs. The other carboxyl-terminal prosequences reported in Table 1 do not show similarity to the propeptides of the RIPs.

The above mentioned homologies suggest the possibility that the propeptides of saporin-6, sap-2 and dianthin-30 may be involved in directing the intracellular traffic of these proteins. All the proteins whose propeptides are reported in Table 1 are inserted into the lumen of the endoplasmic reticulum (ER) and thus enter the secretory pathway. Insertion is due to the presence of an NH$_2$-terminal signal peptide that is co-translationally removed. There is evidence strongly suggesting that, in plant as well as in animal cells, soluble proteins that enter the secretory pathway will be transported from the ER through the Golgi complex and then secreted if they do not carry additional targeting signals (for a review see Vitale and Chrispeels, 1992). These signals may retain proteins in one of the compartments of the pathway or sort them from secretion, probably at the exit of the Golgi complex, to target them to the inner hydrolytic compartments (lysosomes in animal cells and vacuoles in plant cells). The propeptides of barley lectin and chitinase have been shown to be necessary to target these two proteins to the vacuole (Bednarek et al., 1990; Neuhaus et al., 1991). Deletion of these sequences results in secretion of the mutant proteins from transgenic plants. In addition, the two propeptides, when fused to reporter, otherwise secreted proteins, reroute them to the vacuole, indicating that they contain information sufficient for vacuolar targeting (Bednarek and Raikhel, 1991; Neuhaus et al., 1991).

Vacuolar chitinase belongs to a multigene family that encodes proteins that are either secreted or vacuolar (Vitale and Chrispeels, 1992). The secreted chitinase does not have the carboxyterminal propeptide (Payne et al., 1990a). A similar situation appears to occur in the case of glucanase, which is also both vacuolar and secreted: although an involvement at the prosequence of glucanase in vacuolar delivery has not been demonstrated, genes coding for glucanase without the propeptide have been identified (Payne et al., 1990b; van den Bulke et al., 1989; Shinshi et al., 1988). Thus, the observation that the propeptides of saporin-6, sap-2 and dianthin-30 have some homology with those of barley lectin and vacuolar glucanase suggests that the propeptides of the three RIPs may contain information for vacuolar targeting. This hypothesis can be tested by expressing in transgenic plant cells the RIPs, with or without their prosequences, or by fusing the prosequences to reporter proteins.

The presence of a glycosylation site in these propeptides is most probably not a vacuolar targeting signal per se: destruction of the glycosylation site of barley lectin by mutagenesis does not affect vacuolar delivery (Wilkins et al., 1990). Also the presence of a carboxyl-terminal propeptide is not a vacuolar targeting signal per se: the secreted, cell wall-located enzyme polygalacturonidase contains a carboxyl-terminal propeptide (Osteryoung et al., 1990; Table 1). The precise intracellular location of the above mentioned saporins and dianthins is still unknown. A dianthin from Dianthus barbatus and certain Type 1 RIPs from pokeweed have been found in the cell wall matrix of leaf tissues (Frotschl et al., 1990;

Ready et al., 1986). Trichosanthin has been found to be coded by several
genes, each containing or lacking a coding sequence for a carboxyl-terminal
propeptide (Chow et al., 1990; Chow and Piatak, 1990). However, the
carboxyl-terminal propeptides of trichosanthin and of momorcharin, a Type 1
RIP from Momordica charantia (Ho et al., 1991) do not have evident homology
to the propeptides of the saporins, of dianthin or of vacuole-located
proteins (Table 1). Thus, it may be that also saporins and dianthins are
coded by multigene families that code for secreted or vacuolar proteins,
like chitinases and β-glucanases. Although this hypothesis clearly awaits
more data to be confirmed, in this case the double location may also suggest
a potential role of these RIPs as a response to pathogen infection, as has
been postulated for the above mentioned hydrolases. For example, β-1,3-
glucanase from ethylene-treated bean leaves accumulates in vacuoles but is
also found over the middle lamella of the cell wall (Mauch and Staehelin,
1989). This would be supported by the recent finding on the synergistic
inhibition of fungal growth exhibited by one of the barley RIPs with
chitinase and β-glucanase (Leah et al., 1991).

EXPRESSION

       Cloning and expressing a plant toxin gene in micro-organisms is
advantageous because contaminating components conferring toxicity can be
totally eliminated by this approach. Moreover, the assembly of chimaeric
toxins by gene fusion techniques results in a precise peptidic linkage
between the toxic moiety and the ligand. Finally, components of the
conjugate and the conjugate itself can be obtained by cost-effective
fermentation of recombinant micro-organisms.

       Preliminary attempts at expressing recombinant saporin-6 in E. coli
have been performed by fusing the coding region of the gene to that of beta-
galactosidase. Starting from the saporin-6 gene carried by one of the cDNA
clones, an in-frame fusion between the gene coding for beta-galactosidase
and the gene containing most of the coding sequence for saporin-6 was
obtained. The molecular weight of the hybrid beta galactosidase - SO6
protein was found to be, as expected, 145 Kd on SDS-PAGE. The band
migrating at this position on the gel was specifically recognized by an
anti-SO6 antiserum as shown by immunoblot analysis (Lorenzetti et al., 1988).
Purification of the hybrid protein followed by SDS-PAGE of the purified
material revealed a prominent band of the expected molecular weight that was
recognized by the specific anti-saporin-6 antiserum after immunoblotting.

       Progress with plant RIPs as partners for totally recombinant conjugates
has been slower than with their bacterial counterparts, i.e. those obtained
with Pseudomonas exotoxin or diphtheria toxin derivatives (reviewed by
Soria, 1989a). Therefore, the reported E. coli expression of ricin and
abrin A chain, and of a synthetic gene coding for a Mirabilis jalapa Type 1
RIP, is an important preliminary step in this direction (O'Hare et al., 1987;
Habuka et al., 1989). E. coli expression of recombinant saporin-6 has now
been achieved (Prieto et al., 1991), and an active RIP has been obtained
(I. Barthélemy, M.R. Soria and D.A. Lappi, submitted).

CONJUGATION STUDIES

       Reviews on saporin-based immunoconjugates have appeared recently
(Soria, 1989a; Soria et al., 1992). A mitotoxin made of saporin-6, purified
from seeds, linked to recombinant human basic fibroblast growth factor
(bFGF; Bergonzoni et al., 1988; Sarmientos et al, 1988) was recently obtained
by biochemical conjugation using the cross-linker N-succinimidyl-3-(2-

Table 2. Some recent growth factor-based saporin conjugates

Effector	Cytotoxicity	References
Anti-rat nerve growth factor receptor monoclonal antibodies	Superior cervical ganglion neurons (Immunolesioning)	Wiley et al., 1991
Recombinant human basic fibroblast growth factor	Contaminating fibroblasts for pancreatic islet cell purification	Beattie et al., 1990; 1991
" "	Neurons of the CA3 region of the hippocampus	Gonzalez et al., 1991
" "	Smooth muscle cells after balloon catheter injury	Lidner et al., 1991
" "	Cells derived from Dupuytren's contracture	Lappi et al., 1992
" "	Human melanoma cells in nude mice	Beitz et al., 1992
Anti-human Interleukin-2 receptor (CD25) monoclonal antibodies	Activated T lymphocytes	Tazzari et al., 1992a

pyridylthio) propionate (SPDP) and was shown to effectively and selectively kill bFGF receptor-bearing cells with an ID_{50} of 25 pM (Lappi et al., 1989). Similarly powerful conjugates have also been obtained when saporin was conjugated to Ber-H2, an antiCD30 monoclonal antibody, yielding an ID_{50} of less than $5x10^{-13}$ on Hodgkin's derived $CD30^+$ cell lines (Tazzari et al., 1992b).

Receptor-targeted ligand-toxin conjugates (mitotoxins, oncotoxins, hormonotoxins) compete with the respective ligand for receptor binding and, unlike other ligand analogs that behave as competitive antagonists, have the added advantage of irreversibly mediating cell death. Thus, the more powerful mitotoxins offer great potential for opening new areas of investigation into the biology of ligand-receptor interactions, by removing specific cell types in a similar fashion to the work of early physiologists who used organ removal to predict endocrine function (Table 2; reviewed by Lappi and Baird, 1991).

CONCLUSIONS

Molecular strategies for recombinant drug delivery will benefit from accrued knowledge and tools deriving from the explosive growth taking place in the characterization of cellular interactions and of intracellular traffic signals of the type described above (Soria, 1989b). However, detailed knowledge of the behaviour of the individual components of a targeted delivery system, once it is injected in vivo, is still in its infancy. Many disappointments were created in the field of immunotoxins, of liposomes, and of other delivery systems, when promising in vitro systems failed to yield results in vivo. It might be that fixed-type combinations, such as "magic bullets" of even highly sophisticated recombinant manufacture (like totally domain-engineered ligand-toxins or immunotoxins) might not live up to expectations once confronted with the hard facts of in vivo evaluation. Therefore, we should start asking ourselves whether more flexible regimens should be introduced in designing targeting strategies for diagnosis and therapy. Multi-step strategies of the type recently introduced in radioimmunodiagnostics are a good example of improved results deriving from such innovative approaches (Paganelli et al., 1991).

Acknowledgements

This work was supported in part by the Ministero della Sanità - Istituto Superiore di Sanità - IV, V Progetto AIDS 1991, 1992.

REFERENCES

Beattie, G.M., Lappi, D.A., Baird, A. and Hayek, A., 1990, Selective elimination of fibroblasts from pancreatic islet monolayers by basic fibroblast growth factor-saporin mitotoxin, Diabetes, 39:1002.

Beattie, G.M., Lappi, D.A., Baird, A. and Hayek, A., 1991, Functional impact of attachment and purification in the short-term culture of human pancreatic islets, J.Clin.Endocrinol.Metab., 73:93.

Bednarek, S.Y., Wilkins, T.A., Dombrowski, J.E. and Raikhel, N.V., 1990, A carboxyl-terminal propeptide is necessary for proper sorting of barley lectin to vacuoles of tobacco, Plant Cell, 2:1145.

Bednarek, S.Y. and Raikhel, N.V., 1991, The barley lectin carboxyl-terminal propeptide is a vacuolar protein sorting determinant in plants, Plant Cell, 3:1195.

Beitz, J., Davol-Lewis, P., Clark, J., Kato, J., Medina, M., Frackelton, A.R., Lappi, D.A., Baird, A. and Calabresi, P., 1992, Inhibitory effects of the mitotoxin FGF-saporin on human melanoma growth in vitro and in vivo, Cancer Res., 52:227.

Benatti, L., Saccardo, B., Dani, M., Nitti, G., Sassano, M., Lorenzetti, R., Lappi, D.A. and Soria, M., 1989, Nucleotide sequence of cDNA coding for saporin-6, a type-1 ribosome-inactivating protein from Saponaria officinalis, Eur.J.Biochem., 183:465.

Benatti, L., Nitti, G., Solinas, M., Valsasina, B., Vitale, A., Ceriotti, A. and Soria, M.R., 1991, A saporin-6 cDNA containing a precursor sequence coding for a carboxyl-terminal extension, FEBS Lett., 291:285.

Bergonzoni, L., Isacchi, A., Cauet, G., Caccia, P., Sarmientos, P. and Soria, M., 1988, Expression and characterization of recombinant human basic fibroblast growth factor and its molecular variants in E.coli, EMBL Conference "Oncogenes and Growth Control", Heidelberg.

van den Bulcke, M., Bauw, G., Castresana, C., Van Montagu, M. and Vandekerckhove, J., 1989, Characterization of vacuolar and extra-cellular b(1,3)-glucanases of tobacco: evidence for a strictly compartmentalized plant defense system, Proc.Natl.Acad.Sci.USA, 86:2673.

Chow, T.P., Feldman, R.A., Lovett, M. and Piatak, M., 1990, Isolation and DNA sequence of a gene encoding alpha-trichosanthin, a Type I ribosome-inactivating protein, J.Biol.Chem., 265:8670.

Chow, T.P. and Piatak, M., 1990, Genomic cloning of ribosome-inactivating proteins deriving from a multi-gene family in Trichosanthes kirilowii, Abs. 2nd Int. Symp. on Immunotoxins, Orlando.

Collins, E.J., Robertus, J.D., LoPresti, M., Stone, K.L., Williams, K.R., Wu, P., Hwang, K. and Piatak, M., 1990, Primary amino acid sequence of alpha-trichosanthin and molecular models for abrin A-chain and alpha-trichosanthin, J.Biol.Chem., 265:8665.

Fordham-Skelton, A.P., Taylor, P., Hartley, M.R. and Croy, R.R.D., 1991, Characterization of saporin genes: in vitro expression and ribosome inactivation, Mol.Gen.Genet., 229:460.

Frotschl, R., Schonfelder, M., Mundry, K.W. and Adam, G., 1990, Functional studies and subcellular distribution of RIPs from plants with anti-viral activity, Abs. 8th Int. Cong. Virol., Berlin.

Gonzalez, A., Lappi, D.A., Buscaglia, M.L., Carman, L.S., Gage, F.H. and Baird, A., 1991, Basic FGF-SAP mitotoxin in the hippocampus: specific lethal effect on cells expressing the basic FGF receptor, Ann.N.Y. Acad.Sci., in press.

Habuka, N., Murakami, Y., Noma, M., Kudo. T. and Horikoshi, K., 1989, Amino acid sequence of Mirabilis antiviral protein, total synthesis of its gene and expression in E. coli., J.Biol.Chem., 264:6629.

Ho, W.K.K., Liu, S.C., Shaw, P.C., Yeung, H.W., Ng, H.W. and Chan, W.Y., 1991, Cloning of the cDNA of alpha-momorcharin: a ribosome inactivating protein, Biochim.Biophys.Acta, 1088:311.

Houston, L.L., Ramakrishnan, S. and Hermodson, M.A., 1983, Seasonal variations in different forms of pokeweed antiviral protein, a potent inactivator of ribosomes, J.Biol.Chem., 258:9601.

Kornfeld, R. and Kornfeld, S., 1985, Assembly of asparagine-linked oligosaccharides, Ann.Rev.Biochem., 54:631.

Kung, S.S., Kimura, M., and Funatsu, G., 1990, The complete amino acid sequence of antiviral protein from the seeds of pokeweed (Phytolacca americana), Agric.Biol.Chem., 54:3301.

Lamb, F.I., Roberts, L.M. and Lord, J.M., 1985, Nucleotide sequence of cloned cDNA coding for preproricin, Eur.J.Biochem., 148:265.

Lappi, D.A., Esch, F., Barbieri, L., Stirpe, F. and Soria, M., 1985, Characterization of a ribosomal inactivating protein from seeds of Saponaria officinalis (soapworth): immunoreactivities and sequence homologies, Biochem.Biophys.Res.Commun., 129:934.

Lappi, D.A., Martineau, D. and Baird, A., 1989, Biological and chemical characterization of basic FGF-saporin mitotoxin, Biochem.Biophys. Res. Comm., 160:917.

Lappi, D.A. and Baird, A., 1991, Mitotoxins: growth factor-targeted cytotoxic molecules, Progr.Growth Factor Res., 2:223.

Lappi, D.A., Martineau, D., Maher, P.A., Florkiewicz, R.Z., Buscaglia, M., Gonzalez, A.M., Fox, R. and Baird, A., 1992, Basic growth factor in cells derived from Dupuytren's contracture: synthesis, presence and implications for the therapy of the disease, J.Hand Surg., in press.

Leah, R., Tommerup, H., Svendsen, I. and Mundy, J., 1991, Biochemical and molecular characterization of three barley seed proteins with antifungal proteins, J.Biol.Chem., 266:1564.

Legname, G., Bellosta, P., Gromo, G., Modena, D., Keen, J.N., Roberts, L.M. and Lord, J.M., 1991, Nucleotide sequence of cDNA coding for Dianthin 30, a ribosome inactivating protein from Dianthus caryophyllus, Biochim.Biophys.Acta, 1090:119.

Lindner, V., Lappi, D.A., Baird, A., Majack, R.A. and Reidy, M.A., 1991, Role of basic fibroblast growth factor in vascular lesion formation, Circul.Res., 68:106.

Lorenzetti, R., Benatti, L., Dani, M., Lappi, D.A., Saccardo, B.M. and Soria, M., 1988, Nucleotide sequence encoding plant ribosome-inactivating protein, Brit. Pat. appl. n. 8801877.

Maras, B., Ippoliti, R., De Luca, E., Lendaro, E., Bellelli, A., Barra, D., Bossa, F. and Brunori, M., 1990, The amino acid sequence of a ribosome-inactivating protein from Saponaria officinalis seeds, Biochem.Internat., 21:631.

Mauch, F. and Staehelin, L.A., 1989, Functional implications of the subcellular localization of ethylene-induced chitinase and β-1,3 glucanase in bean leaves, Plant Cell, 1:447.

Neuhaus, J.-M., Sticjher, L., Meins, F. and Boller, T.A., 1991, A short C-terminal sequence is necessary and sufficient for targeting of chitinases to the plant vacuole, Proc.Natl.Acad.Sci.USA, 88:10362.

O'Hare, M., Roberts, L.M., Thorpe, P.E., Watson, G.J., Prior, B. and Lord, J.M., 1987, Expression of ricin A chain in E. coli., FEBS Lett., 216:73.

Osteryoung, K.W., Toenjes, K., Hall, B., Winkler, V. and Bennett, A.B., 1990, Analysis of tomato polygalacturonase expression in transgenic tobacco, Plant Cell. 2:1239.

Paganelli, G., Magnani, P., Zito, F., Villa, E., Sudati, F., Lopalco, L., Rossetti, C., Malcovati, M., Chiolerio, F., Seccamani, E., Siccardi, A.G. and Fazio, F., 1991, Three-step monoconal antibody tumor targeting in carcinoembryonic antigen-positive patients, Cancer Res., 51:5960.

Payne, G., Ahl, P., Moyer, M., Harper, A., Beck, J., Meins, F. and Ryals, J., 1990a, Isolation of complementary DNA clones encoding pathogenesis-related proteins P and Q, two acidic chitinases from tobacco, Proc.Natl.Acad.Sci.USA, 87:98.

Payne, G., Ward, E., Gaffney, T., Ahl Goy, P., Moyer, M., Harper, A., Meins Jr., F. and Ryals, J., 1990b, Evidence for a third structural class of β-1,3-glucanase in tobacco, Plant Mol.Biol., 15:797.

Prieto, I., Lappi, D.A., Ong, M., Matsunami, R., Benatti, L., Villares, R., Soria, M., Sarmientos, P. and Baird, A., 1991, Expression and characterization of a basic fibroblast growth factor-saporin-6 fusion protein in E. coli, Ann.N.Y.Acad.Sci., 638:434.

Ready, M.P., Brown, D.T. and Robertus, J.D., 1986, Extracellular localization of pokeweed antiviral protein, Proc.Natl.Acad.Sci.USA, 83:5053.

Sarmientos, P., Isacchi, A., Bergonzoni, L., Cauet, G., Caccia, P. and Soria, M., 1988, Basic fibroblast growth factor: expression, characterization and site-directed mutagenesis of the recombinant molecule, Proc. 34th. Cong. Ital. Soc. Biochem., Padova.

Shinshi, H., Wenzler, H., Neuhaus, J.M., Felix, G., Hofsteenge, J. and Mein, F. Jr., 1988, Evidence for N- and C-terminal processing of a plant defense-related enzyme: primary structure of tobacco prepro-beta-1,3-glucanase, Proc.Nat.Acad.Sci.USA, 85:5541.

Soria, M. and Martini, D., 1987, Recombinant strategies in the search for targeted pharmaceuticals, in: Proc. 4th European Congress on Biotechnology, vol. 4, O.M. Neijssel, R.R. Van der Meer and K.Ch.A.M. Luyben (eds.), Elsevier, Amsterdam.

Soria, M., 1989a, Immunotoxins, ligand-toxin conjugates and molecular targeting, Pharmacol.Res., 21:35.

Soria, M., 1989b, Molecular targeting and delivery: applications of recombinant DNA technology, Biotechnol.Appl.Biochem., 11:527.

Soria, M., 1990, Protein structure and gene organization of saporin-6, a Type 1 ribosome inactivating protein with unusual resistance to proteases, 2nd Int. Symp. on Immunotoxins, Orlando, FLA.

Soria, M., Benatti, L., Lorenzetti, R., Ceriotti, A., Solinas, M., Nitti, G. and Lappi, D.A., 1992, Studies on ribosome inactivating proteins from Saponaria officinalis, in: "Genetically Engineered Toxins", A. Frankel (ed.), Marcel Dekker, New York, in press.

Stirpe, F., Gasperi-Campani, A., Barbieri, L., Falasca, A., Abbondanza, A. and Stevens, W.A., 1983, Ribosome-inactivating proteins from the seeds of Saponaria officinalis L. (soapwort), of Agrostemma githago L. (corn cockle) and Asparagus officinalis L. (asparagus) and from the latex of Hura crepitans L. (sandbox tree), Biochem.J., 216:617.

Stirpe, F. and Barbieri, L., 1986, Ribosome-inactivating proteins up to date, FEBS Lett., 195:1.

Stirpe, F., Barbieri, L., Battelli, M.G., Soria, M. and Lappi, D.A., 1992, Ribosome inactivating proteins from plants: present status and future prospects, Bio/Technol., 10:405.

Tazzari, P.L., Bolognesi, A., De Totero, D., Pileri, S., Conte, S., Wijdenes, J., Hervé, P., Soria, M.R., Stirpe, F. and Gobbi, M., 1992a, BB-10 (AntiCD25) - saporin immunotoxin: a possible tool in graft versus host disease treatment, Transplantation, in press.

Tazzari, P.L., Bolognesi, A., De Totero, D., Falini, B., Lemoli, R.M., Soria, M.R., Pileri, S., Gobbi, M., Stein, H., Flenghi, L., Martelli, M.F. and Stirpe, F., 1992b, BerH2(Anti-CD30) - saporin immunotoxin, a new tool for the treatment of Hodgkin's disease and CD30-positive lymphoma: in vitro evaluation, British J.Haematol., 81:203.

Thorpe, P.E., Brown, A.N., Bremner, J.A.G. Jr., Foxwell, B.M.J. and Stirpe, F., 1985, An immunotoxin composed of monoclonal anti Thy 1.1 antibody and a ribosome-inactivating protein from Saponaria officinalis: potent antitumour effects in vitro and in vivo, J.Natl.Cancer Inst., 75:151.

Vitale, A. and Chrispeels, M.J., 1992, Sorting of proteins to the vacuoles of plant cells, BioEssays, 14:151.

Wiley, R.G., Oeltmann, T.N., and Lappi, D.A., 1991, Immunolesioning: selective destruction of neurons using immunotoxin to rat NGF receptor, Brain Res, 562:149.

Wilkins, T.A., Bednarek, S.Y. and Raikhel. N.V., 1990, Role of propeptide glycan in post-translational processing and transport of barley lectin to vacuoles in transgenic tobacco, Plant Cell, 2:301.

TARGETING WITH IgG AND IMMUNOLIPOSOMES TO CIRCULATING CELLS:

THE 'TARGET CELL DRAGGING' CONCEPT

Daan J.A. Crommelin[1], Pierre A.M. Peeters[2]
and Wynand M.C. Eling[3]

[1]Dept. Pharmaceutics, University of Utrecht
P.O. Box 80.082, The Netherlands
[2]Pharma bio-Research Consultancy BV, Zuidlaren
The Netherlands
[3]Dept. Medical Parasitology, University of Nijmegen
Nijmegen, The Netherlands

INTRODUCTION

In conventional drug targeting approaches attempts are made to accumulate a drug-carrier-homing device combination at the target site. Three important issues should be dealt with appropriately to make this approach successful: (1) access of the combination to the target site, (2) timing of the delivery of the drug and (3) retention of the drug at the target site (Tomlinson, 1987). Most target cells or tissues are localized in a fixed position in the body. The drug-carrier-homing device combination needs to gain access to the target sites, adhere to them and release its load, either inside or in the close proximity of the target. Unfortunately, accumulation of relatively large fractions of the drug at target sites outside the circulation (e.g. solid tumors) after intravenous administration has been found to be more the exception than the rule. This applies both to macromolecular carriers and to particulate carrier systems. Access to target sites present in the blood circulation (e.g. certain endothelial cells and macrophages, and blood clots) is (in principle) less restricted. Successful delivery of drugs to macrophages with particulate systems, such as liposomes laden with immunomodulating agents, antibiotics, antiparasitic or antiviral agents, has been described extensively in reviews (e.g. Emmen and Storm, 1987; Alving, 1988; Nässander et al., 1990; Storm et al., 1991). Huang and co-workers demonstrated that immunoliposomes bearing antibodies with the proper specificity for mouse lung endothelial cells indeed accumulate to a high degree in mouse lungs upon intravenous injection (Hughes et al., 1989).

A different situation is encountered when the target is circulating in the bloodstream. Under the proper conditions attachment of a homing device-carrier combination (such as immunoglobulins) to circulating material can induce uptake of the target-homing device complex by macrophages, mainly those located in the liver (Kupffer cells) and spleen. This resembles a long recognized pathway in the body defense system to dispose of circulating pathogens. Proteins and other macromolecules from the blood bind onto the surface of 'foreign'

Targeting of Drugs 3: The Challenge of Peptides and Proteins
Edited by G. Gregoriadis et al., Plenum Press, New York, 1992

particles and macrophages subsequently act as 'scavenger cells' to remove
the target material from the circulation. Already a decade ago liposomes
carrying digoxin antibodies were used to reduce the concentration of
circulating digoxin in the case of digoxin intoxication (reviewed by
Patel and Ryman, 1981). Attempts to improve antibody mediated imaging by
reducing background blood levels of the imaging antibody using
immunoliposomes were described by a group at the Charing Cross Hospital,
London (Begent et al., 1982; Barratt et al. , 1984). After intravenous
injection of the labelled imaging antibody, long circulating antibody
molecules form a high background signal and jeopardize the quality of the
image. To reduce this background signal, liposomes with immunoglobulins
with an affinity for the primary imaging antibody (liposomally entrapped
second antibody, LESA) are injected after a certain time lapse. The
interaction of the primary, imaging antibody with the original immunoliposomes produces immunoliposomes with 'adhering' antibodies. These
complexes are taken up by macrophages in liver and spleen. As a
consequence the primary antibody disappears rapidly from the blood
circulation and target tissue/blood ratios of the imaging antibody increase
for tissues low in macrophages in direct contact with blood.

In this contribution the concept of 'dragging' of circulating target
cells to macrophages will be described; this 'dragging' is induced by
adhesion of IgG or immunoliposomes onto their surface. In addition, it
deals with the possibilities for subsequent treatment of the 'dragged'
target cells with the drug released from drug laden (immuno)liposomes taken
up inside these macrophages. The concept is schematically depicted in
Fig. 1. Potential target cells are cells circulating in the blood such as
red blood cells (RBC), leucocytes and thrombocytes.

For successful 'dragging' and treatment of target cells the
following issues have to be considered a priori. The target cell must
circulate and expose specific surface structures. Antibodies should be
raised against epitopes of these specific surface structures. Upon
antibody 'adherence', the target cells should be readily phagocytosed by

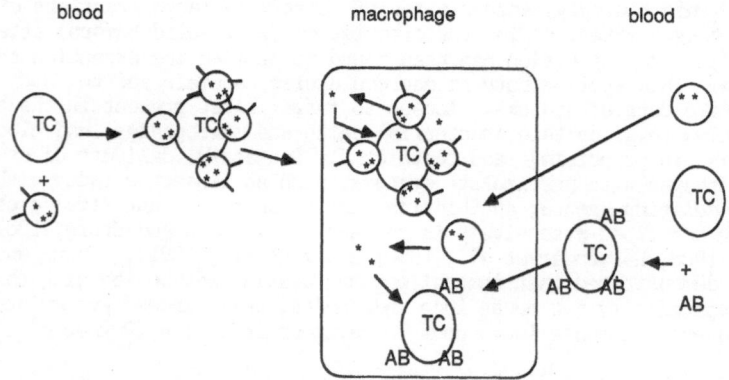

Fig. 1.   Schematic representation of the 'target cell dragging'
          concept. On the left hand side the situation encountered
          with drug (*) laden immunoliposomes, on the right hand side
          the situation where AB and liposomes are administered
          separately.

          AB: antibody, TC: target cell

macrophages. These macrophages should also be able to pick up the relevant drug through (particulate) carrier systems, either prior, simultaneously (in the case of drug laden immunoliposomes) or after capture of the 'dragged' target cell. In Fig. 1 the picture for the two different options is drawn: 1) inject drug laden immunoliposomes, or 2) inject both IgG and, separately (before or after) drug laden liposomes.

Below, both the kinetic aspects of the 'dragging' concept and its therapeutic potential are illustrated in a model system. In the first part the 'dragging' kinetics are dealt with. Cr-labeled mouse RBC, which are not recognized by rat macrophages within the time frame of the experiment, are intravenously injected into the rat. Subsequently, 1) immunoliposomes with anti-mouse RBC Fab fragments, or 2) anti-mouse RBC IgG are injected to convert the Cr-labeled mouse RBC into antibody labelled particles avidly phagocytosed by cells of the mononuclear phagocyte system (MPS). The second part demonstrates the therapeutic potential of the 'dragging' concept. Both chloroquine (CQ)-laden immunoliposomes (option 1, left hand side in Fig. 1), as well as combinations of IgG and 'naked' CQ liposomes (option 2, right hand side in Fig. 1) proved to be superior in treating a Pl. berghei infection in young rats (caused by injection of parasitised mouse RBC (target cells)) to conventional free CQ or naked CQ liposomes.

The critical experiments for the development of the concept were described before in different literature references (Peeters et al., 1988b; 1989b). This manuscript gives the full picture of the 'dragging concept' as developed in our group for the first time. The experimental protocols as used in the key studies (Peeters et al., 1988b, 1989b) are repeated here to allow the reader to fully appreciate the potentials and limitations of the experimental set ups used.

MATERIALS AND METHODS

Chemicals

Egg-phosphatidylcholine (PC), phosphatidylserine (PS), cholesterol, pepsin and dithiothreitol (DTT) were purchased from Sigma Chemicals (St. Louis, MO, USA). Phosphatidylethanolamine (PE) was obtained from Lipid Products (Nutfield, UK). Succinimidylmaleimido-4-(p-phenylbutyrate) (SMPB) was obtained from Pierce Chemical Company (Rockford, USA). Chloroquine-diphosphate met the requirements of the British Pharmacopoeia. Sephadex gels and Protein A-Sepharose CL-4B were purchased from Pharmacia Fine Chemicals (Uppsala, Sweden). Silica gel (70-325 mesh) was obtained from Merck (Darmstadt, FRG) and SE-23 from Servacell (Heidelberg, FRG). $Na_2{}^{51}CrO_4$ was obtained from Amersham, UK.

Preparation of antibody and its fragments

Anti-mouse RBC (anti-mRBC) antibodies were raised in New Zealand white albino rabbits. They received i.v. injections of 10^9 mRBC repeatedly until high haemagglutination (HA) titers in serum were obtained: 2^{16} in a serial twofold dilution test (Salk, 1944). IgG was isolated by Protein A-Sepharose CL-4B affinity chromatography (Goudswaard et al., 1978). Rabbit anti-mRBC $F(ab')_2$ and Fab' fragments were prepared and purified as described earlier (Martin et al., 1981), except that nitrogen instead of argon was used and that the buffer to remove DTT from the Fab'-fragments by PD-10 gelchromatography had a pH of 6.5. The HA titers of the isolated IgG and $F(ab')_2$ were 2^{15} and 2^{12}, respectively, at a protein concentration of 9 mg/ml.

Preparation of CQ containing (immuno)liposomes

MPB-PE was prepared as described earlier (Martin and Papahadjopoulos, 1982). MPB-PE bearing reverse phase evaporation vesicles were prepared by the method of Szoka and Papahadjopoulos (1978) with minor modifications (Peeters et al., 1988b). The lipid composition was chol:PC:PS:MPB-PE 10:9.5:1:0.5. Briefly, the lipids were dissolved in 4 volumes of diethylether (freshly distilled) in a round bottom flask. After addition of glass beads, one volume of CQ solution (225 mg CQ-diphosphate=139 mg CQ base per ml in 100 mM Tris buffer, final pH 4.0) was added and the two phases were emulsified by sonication for 5 min and subsequently mixed on a vortex mixer for one min. Liposomes without CQ were prepared in a similar way with only appropriate buffer components in the hydration solution. Diethylether was removed under reduced pressure in a nitrogen atmosphere with a rotary evaporator at 20°C. The remaining liposome dispersion was subsequently extruded through polycarbonate membranes with 0.4 and 0.2 μm pores (Nucleopore Corp., Pleasanton, USA). After ultracentrifugal sedimentation (to remove non-encapsulated CQ) at 80,000 g during 45 min (3 times) no free CQ was detectable in the supernatant immediately after these washing steps. The final MPB-PE bearing liposome pellet (with CQ) was suspended in isotonic 100 mM sodium acetate solution, pH 7.4 with sodium chloride (lipCQ) and used immediately for coupling to freshly prepared anti-mRBC Fab'-fragments (prepared from anti-mRBC rabbit IgG) as described earlier (e.g. above and Peeters et al., 1988a). Fab'-liposomes (Fab'-lip) with either specific anti-mRBC Fab' or control Fab' (prepared from a feline leukemia virus (FeLV) monoconal IgG_1) with (Fab'-lipCQ) or without encapsulated CQ were prepared in a similar way.

Coupling of Fab' fragments to the MPB-PE bearing reverse phase evaporation vesicles

Freshly prepared MPB-PE bearing reverse phase evaporation vesicles (final concentration 3.5 μmole phospholipid/ml in deoxygenated buffer were mixed with freshly prepared Fab'fragments (final concentration 0.3 mg/ml). The coupling reaction was carried out in a nitrogen atmosphere under stirring at room temperature for 90 min. Fab'-MPB-PE bearing reverse phase evaporation vesicles (Fab'-liposomes) were separated from unconjugated Fab'fragments by ultacentrifugal sedimentation of the Fab'-liposomes at 80,000 g during 45 min in a Beckman TY-65 rotor (Peeters et al., 1988a).

CQ determination

Total amount of CQ (expressed as CQ-base in liposome suspensions) was determined spectrophotometrically at 341 nm, pH 1. Free CQ (CQ leakage) was determined spectrofluorimetrically at pH 9.3, excitation and emission wavelength 330 nm and 383 nm, respectively, and expressed as percentage of the total CQ content as described earlier (Peeters et al., 1989a).

Animals and mRBC used in the blood clearance and tissue distribution studies

Closed colony bred male Swiss mice (15-18 g) and male Wistar rats (140-170 g) were obtained from colonies of the animal facility of the University of Nijmegen. mRBC (from male Swiss mice) were purified by

applying the heparinized whole blood onto a column containing three
volumes of Sephadex G-150 Superfine and one volume of SE-23 to remove
white blood cells (Eling, 1977).

In vivo experiments assessing blood clearance and tissue distribution of labeled mRBC

mRBC (10^9/ml buffer) were labelled with $Na_2^{51}CrO_4$ (about $10\,\mu$
Ci/5×10^8 mRBC) during 60 min at 37 °C, washed twice, resuspended in
buffer and used immediately. In the rat experiments Cr-mRBC (1.5×10^9;
equivalent to about 450,000 cpm per rat) were injected i.v. in the tail
vein of the rat, followed by the indicated amount of Fab'- liposomes,
liposomes, IgG, F(ab')$_2$ or 0.5 ml buffer 30 to 45 min later. At fixed
time intervals, blood samples (50 µl) were taken from the tail vein into
heparinized glass capillaries and the blood was transferred to
scintillation vials and counted for ^{51}Cr radioactivity. Total injected
radioactivity (=100%) was determined by counting a sample of the injected
material. The proportion of radioactivity in the circulation at the
indicated time points was calculated assuming a total blood volume of 6%
with respect to the body weight (Mitruka and Rawnsley, 1977). At the
indicated time points the animals were killed and radioactivity in liver,
spleen and lungs was determined. Total recovery in the results section
refers to the sum of radioactivity in blood, spleen, liver and lungs.
Each data point is the mean ± SD of 3 animals, unless otherwise stated.

Animals used to assess therapeutic efficacy

Outbred male Swiss mice (6 weeks old, 25 g) and male Wistar rats (4-5
weeks old, 50-70 g; young rats were used because of their high
sensitivity to the infection) were obtained from colonies of the animal
facility of the University of Nijmegen. they were kept in plastic cages
and received standard food (RMH, Hope Farms) and water ad libitum.

Parasite

P. berghei (strain K173) was maintained by weekly intraperitoneal
(i.p.) inoculation of 10^5 p (parasitised)-mRBC from infected donor mice
into healthy mice of the same strain and sex. Parasitemia was determined
from blood smears, made from tail blood stained with May Grunwald-
Giemsa's solutions.

Synchronization of the infection

Synchronization was performed according to the methods described by
Mons et al. (1983) and Janse et al. (1984) with minor modifications.
Infected mouse reticulocytes (p-mRETS) used for synchronization were
obtained from 2 Swiss mice made anaemic by bleeding from the retro-
orbital plexus (200-250 µl of blood) one day before and one day after
infection with 10^5 p-mRBC obtained from a passage mouse on day 7 after
infection. Five days after infection the parasitemia was 8-10% with 60%
of the parasites in reticulocytes (p-mRETS). Blood of the infected
animals was taken by cardiac puncture under sterile conditions,
supplemented with heparin (5 U/ml) and diluted in complete medium (RPMI
1640, buffered with 20 mM Herpes and 10 mM sodium hydrogencarbonate and
supplemented with 10% FCS, 2 mM L-glutamin and penicillin (100 U/ml)).
After centrifugation (1500 g, 5 min) the buffy coat was removed, the
pelleted cells washed with complete medium and diluted with normal mRBC
(1:1, obtained in a similar way) in a final volume of 50 ml. This
suspension was cultivated in an erlenmeyer flask (500 ml) in a shaking
apparatus at 37 °C in an atmosphere of 85% nitrogen, 10% oxygen and 5%
carbon dioxide. After 17 h of cultivation the majority of the parasites

had matured to the schizont stage (microscopic analysis). The
synchronized p-mRETS (syn-p-mRETS) were separated by density
centrifugation (1600 g, 20 min) on a cushion of Nycodenz (55% v/v). The
interphase was collected and washed with complete medium. The cells were
counted and stored on ice. 5×10^7 of these mature schizonts were mixed
with $2-3 \times 10^8$ mRETS and immediately injected (i.v.) in a normocytic
mouse (the erythropoiesis of this mouse was inhibited by the i.p.
injection of 0.5 ml heparinized normal whole blood 7 and 4 days
previously). The mRETS that were mixed with the syn-p-mRETS were
obtained from mice that were hyperbled from the retro-orbital-plexus
(200-250 µl) 5 and 2 days previously. These mRETS were isolated by
density centrifugation on a Percoll cushion as described above. 17 h
after inoculation of syn-p-mRETS and mRETS, blood was taken and the syn-
p-mRETS were isolated as described above. Microscopic analysis revealed
that over 90% of the parasites were in the throphozoite stage and in
reticulocytes (syn-p-mRETS). These syn-p-mRETS were stored on ice until
they were injected.

Therapeutic studies to cure rats infected with synchronized parasitized mRBC

10^5 syn-p-mRETS were injected into the tail vein of rats in a total
volume of 0.2 ml. Anti-mRBC Fab'-lipCQ, control Fab'-lipCQ, or lip CQ
(in a dose of 0.6 mg CQ per rat), or anti-mRBC Fab'-lip (no CQ) or buffer
were injected in a total volume of 0.2 ml into the tail vein 10 min

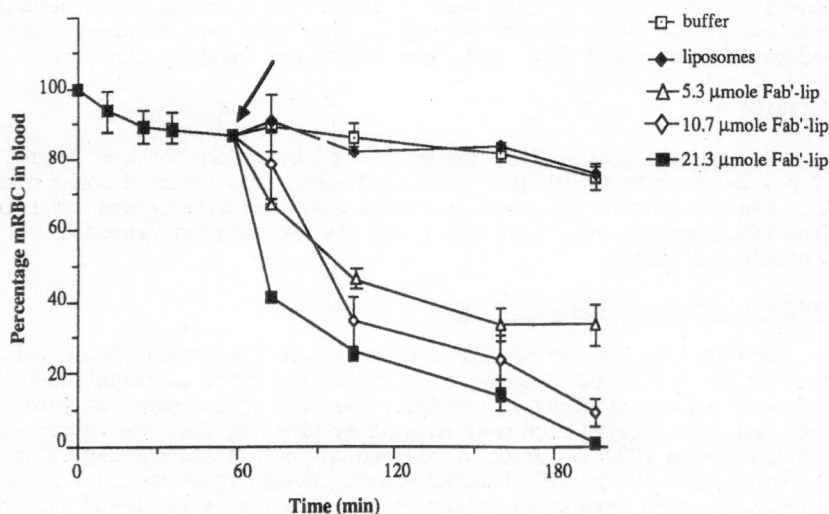

Fig. 2.   Blood level of Cr-mRBC in rats after i.v. administration of
buffer, liposomes (21.1 µmole phospholipid per rat), or
Fab'-liposomes at three doses: 5.3, 10.7 and 21.3 µmole
phospholipid per rat, 50 min (arrow) after injection of Cr-
mRBC (1.5×10^9 per rat). Fab'-liposomes had a ratio of
30.6 µg Fab'/µmole phospholipid and a mean diameter of
0.38 µm, while liposomes had a mean diameter of 0.28 µm.
In all cases a total volume of 0.5 ml was injected. Each
data point represents the mean of at least 3 +/- SD. Small
SD are not shown.

Table 1. Distribution of Cr-mRBC in rats over various organs, blood
and total recovery of ^{51}Cr-label after i.v. administration
of various doses of Fab'-liposomes and controls

Treatment	Blood	Spleen	Liver	Lungs	Total
Buffer	74.9 ± 3.9	8.2 ± 0.2	11.5 ± 0.3	4.4 ± 0.6	99.0 ± 2.0
Liposomes	76.0 ± 2.0	4.1 ± 1.1	9.9 ± 1.0	2.0 ± 0.2	91.9 ± 2.3
5.3 µmole Fab'-lip	33.5 ± 5.5	39.1 ± 2.0	16.4 ± 5.4	2.1 ± 0.2	90.9 ± 1.7
10.7 µmole Fab'-lip	9.0 ± 4.0	53.4 ± 8.4	29.5 ± 9.6	1.4 ± 0.3	93.4 ± 13.8
21.3 µmole Fab'-lip	1.0 ± 0.2	53.0 ± 7.0	31.5 ± 9.8	1.2 ± 0.2	86.1 ± 2.0

Distribution of Cr-mRBC in rats over spleen, liver, lungs, blood and total recovery of ^{51}Cr-label
2.5 h after i.v. administration of buffer, lipsomes (21.1 µmole phospholipid per rat) and Fab'-
lipsomes at three dosage levels: 5.3, 10.7 abd 21.3 µmole phospholipid per rat. Each data point
(% injected Cr) represents the mean of 3 rats ± SD. For further details on lipsomes or Fab'-
lipsomes: see legends of Fig. 2.

later. In cases where the combination of anti-mRBCIgG/lipCQ was tested
the two components were injected separately 10 min after injection of the
syn-p-mRETS. The dose of anti-mRBC IgG was 140 µg per rat (unless
otherwise indicated) which was similar (on a protein basis) to the amount
of Fab'-lipCQ (25 ± 5 µg Fab'/µmole phospholipid; injected dose 5 µmole
per rat). Mortality was recorded.

Other methods used

    Protein was determined by the method of Peterson (1977). Lipid
phosphate was estimated by the colorimetric method of Fiske and SubbaRow
(1925). The quality of the phospholipids and synthesized MPB-PE was
tested by a TLC method (Heath et al., 1981). Particle size analysis was
performed by dynamic light scattering with a Malvern PCS 100 SM (Malvern
Ltd., Worcestershire, UK) equipped with a particle analyzer processor
(Model 7027) and a 27 mWatt helium/neon laser (NEC Corp., Tokyo, Japan).

RESULTS

Distribution of ^{51}Cr-labelled mRBC in rats after targeting of Fab'-
liposomes to mRBC

    A common problem in studying the interaction of immunoliposomes with
their target cells in the blood compartment is that in the animal models
used until now these, sometimes modified, target cells are rapidly
eliminated from the blood compartment; the body recognises them as
'foreign body like' structures even in the absence of a homing device and
removes them rapidly (Wolff and Gregoriadis, 1984; Laakso et al., 1986).
Therefore, there is a need for an in vivo model in which a large fraction
of the target cells (80-90% of the injected dose) will circulate over a
period of several hours. The present study provides such a model: Cr-
labeled mouse RBC intravenously injected into rats are the target cells.
The clearance of Cr-mRBC in rats is slow (Fig. 2). After 3 h 80% of the
injected amount of Cr-mRBC is still circulating. The clearance can be

enhanced considerably when Fab'-liposomes carrying the appropriate homing device (Fab' fragments) are injected. Increasing the dose of Fab'-liposomes from 5.3 to 21.3 µmole phospholipid enhanced the blood clearance of mRBC in a dose dependent way (Fig. 2). Liver and spleen are important organs for the uptake of Cr-mRBC-Fab'-liposome complexes (Table 1). A reduction of total recovery is observed when the highest dose (21.3 µmole Fab'-liposomes) is injected (Table 1). Liposomes alone do not affect the clearance of mRBC (Fig. 2).

Distribution of ^{51}Cr-labelled mRBC in rats after targeting of IgG to mRBC

In the same model, the effect of specific IgG administration (rabbit anti-mouse RBC antibodies; doses ranging from 10 µg to 360 ug IgG) on the Cr-mRBC elimination (Fig. 3) and subsequent uptake into liver and spleen (Table 2) was studied. Again elimination of Cr-mRBC from the blood stream, and subsequent spleen and liver uptake were dose dependent. At relatively low doses (30 µg IgG per rat) mainly spleen uptake occurred. However, at doses of 90 µg IgG or higher, Cr-mRBC were preferentially cleared by the liver while the relative spleen uptake tended to decrease (Table 2).

Clearance and organ distribution of Cr-mRBC in rats after injection with Fab'-lip: effect of Fab' density on liposomes

In order to study the effect of the Fab' molecule density on liposomes on clearance and organ distribution, immunoliposomes were prepared with lower densities, e.g. 15.3, 5.5 and 1.5 µg/µmole phospholipid. This was achieved by lowering the Fab' concentration in

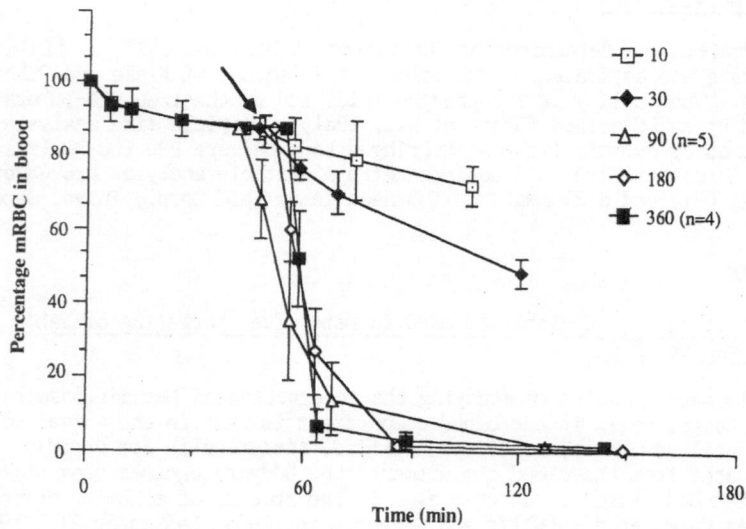

Fig. 3. Blood level of Cr-mRBC in rats after i.v. administration of different IgG doses (10, 30, 90, 180 and 360 µg IgG per rat, approximately 50 min (arrow) after injection of Cr-mRBC (1.5x10^9 per rat) in a total volume of 0.5 ml. Each data point represents the mean of 3 rats (unless otherwise indicated)±SD. Small SD are not shown.

Table 2. Distribution of Cr-mRBC in rats over various organs, blood and total recovery of ^{51}Cr-label after iv. administration of IgG

IgG (µg)	Blood	Spleen	Liver	Lungs	Total
10	69.9 ± 8.5	6.3 ± 1.4	7.6 ± 0.7	1.2 ± 0.1	84.9 ± 10.3
30	48.2 ± 4.0	27.4 ± 6.2	8.2 ± 0.4	0.8 ± 0.6	84.8 ± 2.4
90 (n=5)	1.2 ± 1.4	19.6 ± 6.8	59.5 ± 9.0	0.8 ± 0.7	81.6 ± 6.7
180	0.7 ± 0.7	19.2 ± 3.4	59.8 ± 2.7	1.8 ± 0.2	81.5 ± 6.3
360 (n=4)	1.5 ± 1.3	13.0 ± 2.5	70.4 ± 6.0	1.2 ± 0.8	86.4 ± 6.3

Distribution of Cr-mRBC over spleen, liver, lungs, blood and total recovery of ^{51}Cr-label 1.5 h after i.v. administration of the indicated IgG doses per rat in a total volume of 0.5 ml. Each data point (% injected Cr) represents the mean of 3 rats (unless otherwise indicated) ± SD.

the coupling reaction medium to 0.25, 0.1 and 0.03 mg/ml, respectively. The results of experiments with these immunoliposomes and control preparations are presented in Figs. 4 and 5. The clearance curves and tissue distributions obtained after treatment with liposomes or control immunoliposomes (anti-FeLV Fab'-lip; results not shown) were essentially the same as for specific (anti-mRBC) Fab'-lip with the lowest ratio (1.5 µg Fab'/ mole phospholipid) (Figs. 4 and 5). Increasing the anti-mRBC Fab' density resulted in a faster elimination of Cr-mRBC (Fig. 4) and enhanced localization of the ^{51}Cr label in the liver and particularly in the spleen (Fig. 5).

Effect of administration of Fab'-lipCQ or IgG/lipCQ on an intravenous infection of rats with synchronized parasitized mRETS

The IgG and Fab'-fragments used in these experiments are directed against mRBC membranes and are supposed to be non-reactive to parasite molecules. Syn-p-mRETS used in the following experiments exhibited >90% trophozoites (mRBC form) and no schizonts or merozoites (microscopic analysis). The results of two independent experiments using the syn-p-mRETS are summarized in Table 3. These results indicate that the therapeutic efficacy of anti-mRBC Fab'-lipCQ and the combination of anti-mRBC IgG and 'naked' lipCQ was by far superior to other regimens tested: free CQ, lipCQ or buffer.

DISCUSSION

The results presented in the previous section clearly demonstrate different aspects of the 'target cell dragging' concept: the rapid and (under properly chosen conditions) almost quantitative removal of the circulating target cells from the blood. Probably, the 'dragged' target cells accumulate in the macrophages in contact with the circulation as accumulation occurs in MPS rich organs such as liver (Kupffer cells) and spleen. The tissue distribution of the labeled target cells is assumed to be directly related to the uptake of the Cr-label. Uptake data are presented as the fraction of initial dose per organ and are not corrected

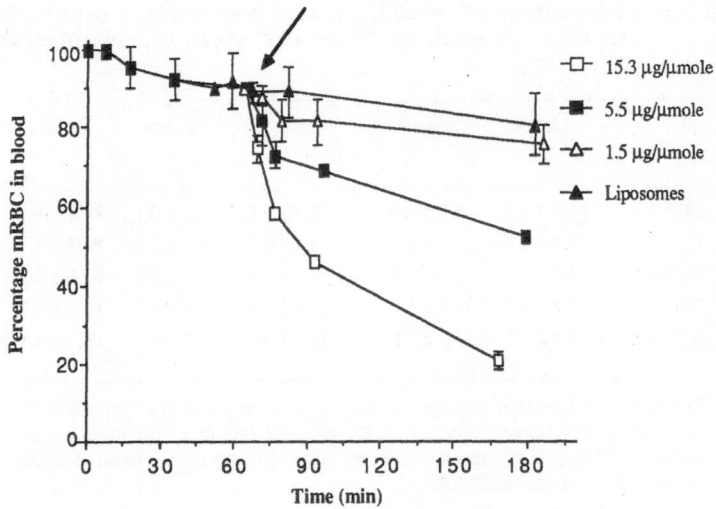

Fig. 4. Blood level of Cr-mRBC (1.5×10^9) in rats after i.v. administration of liposomes or Fab'-lip with different liposomal Fab' densities. In all cases a total amount of 10 μmole phospholipid was injected in a total volume of 0.5 ml 1 h after injection of Cr-mRBC (arrow). Each data point represents the mean ±SD of four animals. Small SD are not shown.

for the organ weight; this implies that spleen uptake efficiency per g tissue can be much higher than liver uptake efficiency, if the data are normalised for tissue weight. The 'dragging' capacity strongly depended on the experimental conditions. Within the range tested an increase in the Fab' fragment density on the liposomes made the target cell removal more efficient. The dragging effect with specific IgG was clearly concentration dependent as well. Interestingly, the ratio liver/spleen uptake depended on the 'dragging agents': with immunoliposomes there was an indication for a preference for uptake by the spleen; with IgG, liver uptake dominated for IgG doses of 90 μg or higher. The underlying mechanism for this preferential uptake by either liver or spleen remains to be determined.

In the murine malaria model used, both 'target cell dragging' protocols that were tested, the CQ laden immunoliposomes directed against parasitized mouse RBC and the combination of anti-mouse RBC IgG plus separately injected 'naked' CQ laden liposomes, proved to be therapeutically superior to conventional CQ therapy or the use of 'naked' CQ liposomes alone. The mechanism behind this improved therapeutic potential has been schematically depicted in Figure 1. The combination of IgG and 'naked' liposomes, injected separately, tended to be even more effective than the technologically much more sophisticated CQ-immunoliposomes and might therefore become the preferred protocol in the animal model used.

Polyclonal rabbit-anti-mouse RBC antibodies were used. They allowed the demonstration of the feasibility of the 'target cell dragging' concept. However, monoclonal antibodies are preferred over polyclonal antibodies, because of their superior pharmaceutical properties (purity,

Table 3. Effect of Fab'-lipCQ or IgG/lipCCQ administration on the development of a patency after intravenous infection with synchronized parasitized mRETS

Treatment	Number of rats radically cured[1]
anti-mRBC Fab'-lipCQ[2]	2 (6)
anti-mRBC IgG/lipCQ[2]	4 (4)
CQ	0 (4)
lipCQ	0 (8)
buffer	0 (8)

[1] No parasites were detected during a 28 days period in rats infected with 10^5 p-mRETS (i.v.).
[2] CQ was given in a dose of 0.6 mg CQ per rat. IgG dose was 140 µg per rat in the indicated treatment.

reproducibility, characterization possibilities and availability in 'gram quantities'). With monoclonal antibodies information should be collected to optimize the therapeutic response. Issues that need further attention are the drug type and dose, dose of homing device and carrier involved, administration protocol, surface density of the Fab' fragments (for the immunoliposomes), IgG class and subtype, safety of the concept and, in particular, the immunogenicity of the different protocols.

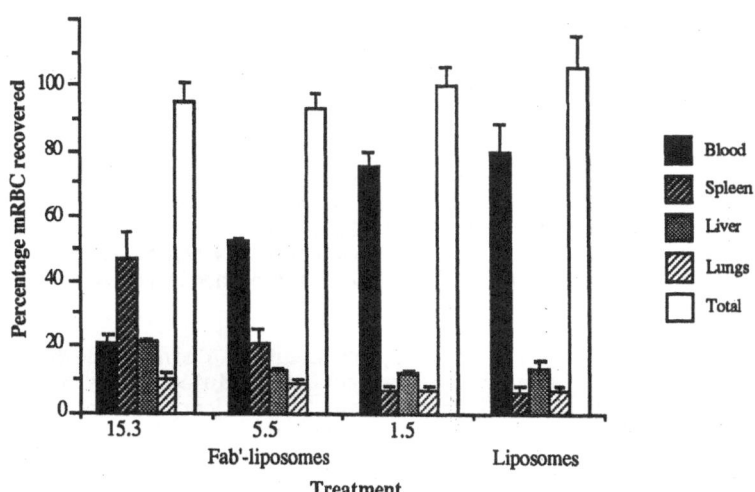

Fig. 5. Recovery of Cr-mRBC in spleen, liver, lungs, blood and their sum. Animals were killed, two h after i.v. administration of liposomes or anti-mRBC Fab'-lip with different liposomal Fab' densities (indicated as µg Fab'/µmole phospholipid). Presented data are the mean ±SD from four animals.

The immunoliposome or IgG-directed 'dragged' targets as well as 'naked' liposomes are presumably phagocytosed by macrophages. This means that the content of the liposome is delivered into the cell. This internalization route can deliver active agents into the cell, that would never cross the cytoplasmic membrane in the 'free' form. The use of the so called liposome dependent drugs (Heath et al., 1985a,b) might show advantages over drugs used in conventional therapy, which are selected, on the basis of their pharmacokinetic profile in the 'free' (not carrier bound) form.

As pointed out in the Introduction, the 'dragging' concept can only be successful, if the target cell can be selectively identified among other cells circulating in the blood. (Monoconal) antibodies or other selecting agents must therefore be available to identify the target cells and to induce the 'dragging' effect. Such monoclonal antibodies recognizing only parasitized erythrocytes are already available for human malaria (Anders et al., 1989). The murine malaria model used in this study should be considered only as a model to demonstrate the potential therapeutic advantage of the 'target cell dragging' concept. Other potential candidate cells or structures for 'dragging' are lymphocytes (e.g. HIV infected cells) and circulating pathogens, which are difficult to treat by conventional therapy.

In some cases the destruction machinery of the macrophage itself might already be sufficient to deal with the 'dragged' pathogen involved. In the case of Leishmania parasites, however, a drug is needed to bring about a therapeutically beneficial effect. In the murine malaria model described in this contribution neither antibodies against the target cells, nor free CQ, nor lipCQ alone were therapeutically effective. A combination of antibodies/immunoliposomes and liposome associated CQ was needed for an effective therapy. Erythrocytic forms of the malaria parasites do not survive outside their host cell, and are readily broken down, once being inside macrophages (Garnham, 1966). It is unclear why antibody dependent dragging of infected erythrocytes alone is not sufficient, whereas the combination with CQ treatment is. An important aspect of the model system (elimination of parasite-infected mRBC in rats) is that free parasites (e.g. merozoites) are not dragged to the macrophages. Any free parasite that escapes the dragging principle can enter a rat RBC and is no longer attacked by the dragging antibody. This explains why treatment is more efficient in stage-synchronized than in asynchronized infected cells (Peeters et al., 1989b). Such infected cells, however, are still sensitive to CQ treatment. Elimination of parasites by chloroquine depends on the dose of CQ used and on the degree of the parasitemia (Peeters et al., 1989b). Thus the escape of the dragging treatment by a minimal number of parasites can be reversed by a single dose of CQ, whereas a higher parasite load needs repeated CQ treatments.

Whether escape from the dragging treatment of a minimal number of parasites is a technical problem (preparations of infected erythrocytes contain a certain number of free parasites and one parasite is enough to start an infection) or relates to escape of parasites after being dragged to the macrophages is not clear, although the latter possibility is considered less likely.

Candidate drugs for liposome encapsulation, which are chosen to exert their therapeutic action in macrophages should be resistant to lysosomal enzymes. This excludes most peptide and protein drugs unless early escape from the endosomes to the cytoplasma after phagocytosis will be induced by using pH sensitive liposomes, which follow the entry through the endocytic pathway of certain envelope viruses: they become fusogenic at the low pH of the endosome (Chu et al., 1990).

REFERENCES

Alving, C.R., 1988, Macrophages as targets for delivery of liposome-encapsulated antimicrobial agents, Adv.Drug Del.Rev., 2:107.

Anders, R.F., Smythe, J.A., Barzaga, N.G., Forsyth, K.P., Brown, H.J., Crewther, P.E., Thomas, L.M., Coppel, R.L., Culvenor, J.G. and Brown, G.V., 1989, in: "New Strategies in Parasitology". K.P.W.J. McAdam, ed., Churchill Livingstone, Edinburgh.

Barratt, G.M., Ryman, B.E., Chester, K.A. and Begent, R.H.J., 1984, Liposomes as aids to tumour detection, Biochem.Soc.Trans., 12:48.

Begent, R.H.J., Green, A.J., Bagshawe, K.D., Jones, B.E., Keep, P.A., Searle, F., Jewskes, R.F., Barratt, G.M. and Ryman, B.E., 1982, Liposomally entrapped second antibody improves tumour imaging with radiolabelled (first) antitumour antibody, Lancet, ii:730.

Chu, C., Dykstra, J., Lai, M., Hong, K. and Szoka, F., 1990, Efficiency of cytoplasmic delivery by pH-sensitive liposomes to cells in culture, Pharm.Res., 7:824.

Eling, W.M.C., 1977, Separation of parasitized erythrocytes from Plasmodium berghei infected mouse blood, Trans.Roy.Soc.Trop.Med.Hyg., 71:267.

Emmen, F. and Storm, G., 1987, Liposomes in treatment of infectious diseases, Pharm.Weekbl.Sci.Ed., 9:162.

Fiske, C. and SubbaRow, Y., 1925, The colorimetric determination of phosphorus, J.Biol.Chem., 66:375.

Garnham, P.C.C., 1966, in: "Malaria parasites and Other Haemosporidia", Blackwell Scientific Publications, Oxford.

Goudswaard, J., Van Der Donk, J.A., Noordzij, A., Van Dam, R.H. and Vaerman, J-P., 1978, Protein A reactivity of various mammalian immunoglobulins, Scand.J.Immunol., 8:21.

Heath, T.D., Macher, B.A. and Papahadjopoulos, D., 1981, Covalent attachment of immunoglobulins to liposomes via glycosphingolipid, Biochim.Biophys.Acta, 640:66.

Heath, T.D., Lopez, N.G. and Papahadjopoulos, D., 1985, The effect of liposome size and surface charge on liposome-mediated delivery of methotrexate-gamma-aspartate to cells, Biochim.Biophys.Acta, 820:74.

Heath, T.D., Lopez, N.G., Stern, W.H. and Papahadjopoulos, D., 1985, 5-Fluoroorotate: a new liposome-dependent cytotoxic agent, FEBS Lett., 187:73.

Hughes, B.J., Kennel, S., Lee, R. and Huang, L., 1989, Monoclonal antibody targeting of liposomes to mouse lung in vivo, Cancer Res., 49:6214.

Janse, C.J., Mons, B., Croon, J.J.A.B. and van der Kaay, H.J., 1984, Long-term in vitro cultures of Plasmodium berghei and preliminary observations on gametocytogenesis, Int.J.Parasitol., 14:317.

Laakso, T., Andersson, J., Artursson, P., Edman, P. and Sjöholm, I., 1986, Acrylic microspheres in vivo. X. Elimination of circulating cells by active targeting using specific monoclonal antibodies bound to microparticles, Life Sci., 38:183.

Martin, F.J., Hubbel, W.L. and Papahadjopoulos, D., 1981, Immunospecific targeting of liposomes to cells: a novel and efficient method for covalent attachment of Fab' fragments via disulfide bonds, Biochemistry, 20:4229.

Martin, F.J. and Papahadjopoulose, D., 1982, Irreversible coupling of immunoglobulin fragments to reformed vesicles, J.Biol.Chem., 257:286.

Mitruka, B.M. and Rawmsley, H.M., 1977, "Clinical, Biochemical and Haematological References in Normal Experimental Animals", Mason Publ. USA, Inc., New York.

Mons, B., Janse, C.J., Croon, J.J.A.B. and Van der Kaay, H.J., 1983, In vitro culture of Plasmodium berghei using a new suspension system, Int.J.Parasitol., 13:213.

Nässander, U.K., Storm, G., Peeters, P.A.M. and Crommelin, D.J.A., 1990, Liposomes, in: "Biodegradable Polymers as Drug Delivery Systems", M. Chasin and R. Langer, R., eds., Marcel Dekker, NY, 1990.

Patel, H.M. and Ryman, B.E., 1981, Systemic and oral administration of liposomes, in: "Liposomes: From Physical Structure to Therapeutic Applications", C.G. Knight, ed., Elsevier/North Holland, Amsterdam.

Peeters, P.A.M., Claessens, C.A.M., Eling, W.M.C. and Crommelin D.J.A. 1988a, Immunospecific targeting of liposomes to erythrocytes, Biochem.Pharmacol., 37:2215.

Peeters, P.A.M., Oussoren, C., Eling, W.M.C and Crommelin D.J.A., 1988b, Immunospecific targeting of immunoliposomes, F(ab')2 and IgG to red blood cells in vivo, Biochim.Biophys.Acta, 943:137.

Peeters, P.A.M., Huiskamp, C.W.E.M., Eling, W.M.C. and Crommelin, D.J.A., 1989a, Chloroquine containing liposomes in the chemotherapy of murine malaria, Parasitology, 98:381.

Peeters, P.A.M., Brunink, B.G., Eling, W.M.C. and Crommelin, D.J.A., 1989b, Therapeutic effect of chloroquine (CQ) containing immuno-liposomes, or combinations of antibodies and CQ or liposomal CQ in rats infected with Plasmodium berghei parasitized mouse red blood cells, Biochim.Biophys.Acta, 981:269.

Peterson, G.L., 1977, A simplification of the protein assay of Lowrey et al., which is more generally applicable, Anal.Biochem., 83:346.

Salk, J.E., 1944, Simplified procedure for titrating hemagglutinating capacity of influenza-virus and corresponding antibody, J.Immunol., 49:87.

Storm, G., Wilms, H.P. and Crommelin, D.J.A., 1991, Liposomes and biotherapeutics, Biotherapy, 3:25.

Szoka, F. and Papahadjopoulos, D., 1978, Procedure for preparation of liposomes with large internal aqueous space and high capture by reverse-phase evaporation, Proc.Natl.Acad.Sci.USA, 75:4194.

Tomlinson, E., 1987, Theory and practice of site-specific drug delivery, Adv.Drug Deliv.Rev., 1:87.

Wolff, B. Gregoriadis, G., 1984, The use of monoclonal anti-Thy$_1$ IgG$_1$ for the targeting of liposomes to AKR-A cells in vitro and in vivo, Biochim.Biophys.Acta, 802:259.

LIPOSOME AND IMMUNOLIPOSOME MEDIATED DELIVERY OF PROTEINS AND PEPTIDES

Leaf Huang and Fan Zhou

Department of Pharmacology
University of Pittsburgh
School of Medicine
Pittsburgh, PA 15261, USA

INTRODUCTION

Proteins and peptides have emerged to become a class of important pharmaceuticals largely because of technological advances in the recombinant DNA and automated peptide synthesis. Efficient delivery of these new drugs to the appropriate target site presents itself as a serious challenge. Proteins and peptides can be categorized into two groups: those acting by binding to the cell surface receptor and those acting in the cytoplasm. Strategies for delivering the two types of proteins and peptides are quite different. Liposomes and immunoliposomes (those with antibody covalently attached on the surface) have been used to deliver both proteins and peptides, but the specific formulation varies considerably depending on which group the protein or peptide belongs to.

PROTEINS WITH ACTION ON THE CELL SURFACE

Entrapment of proteins and peptides in liposomes usually provides a protection of these drugs from inactivation by extracellular proteases and peptidases (Fishman and Citri, 1975). In some cases the protein or peptide is also stabilized by the incorporation in liposomes (Manning et al, 1989). However, these macromolecules have to be released from liposomes once they have arrived at the target site. Free diffusion across the liposomal membranes is usually too slow to achieve the desired bioavailability for these macromolecules. A controlled or triggered destabilization of the liposome membrane would result in a quantitative release of the encapsulated drug at the site. The target-sensitive immunoliposomes are designed to serve this purpose.

TARGET-SENSITIVE IMMUNOLIPOSOMES

We have taken advantage of the polymorphic property of unsaturated phosphatidylethanolamine (PE) to design the new immunoliposomes. The equilibrium state of PE at the physiological conditions is that of an inverted micellar phase (H_{II} phase) (Cullis and de Kruijff, 1979). However, the bilayer L phase of PE can be stabilized by mixing PE with a

Targeting of Drugs 3: The Challenge of Peptides and Proteins
Edited by G. Gregoriadis et al., Plenum Press, New York, 1992

45

hydrophilic, often charged, amphiphile. This is mainly because the increased interfacial hydration prevents close bilayer contacts among PE bilayers which is required for the transition to the H_{II} phase (Ellens et al, 1986). Many hydrophilic amphiphiles have been used to stabilize PE liposomes. IgG antibody molecules which have been derivatized with long-chain fatty acids are one of the stabilizers although the stabilization activity is relatively weak (Ho et al, 1986a). Immunoliposomes containing dioleoyl or other unsaturated PE and approximately 1% acylated IgG antibody spontaneously and rapidly destabilize when they bind to a multivalent target which expresses many copies of the antigen molecules (Ho et al, 1986b). Liposome destabilization, which leads to the content release, is immunospecific and can be blocked by pretreatment of the target with an excess of free antibody (Ho et al, 1987b). These target-sensitive immunoliposomes have been used for the construction of a homogenous liposome immunoassay (LIA) for the herpes simplex virus (HSV). They have also been used in an in vitro model system as a target specific drug delivery system. Monoclonal antibody to the gD glycoprotein of HSV was used to construct the target-sensitive immunoliposomes encapsulating antiviral drugs, acyclovir or AraC (Ho et al, 1987a). Mouse L929 cells which had been infected with HSV at low multiplicity were treated with these immunoliposomes and the virus titer of the culture was assayed 42 hours later. Furthermore, the cytotoxicity of the treatment was also assayed by the inhibition of $[^3H]$-thymidine incorporation into the cellular DNA. The results of these experiments showed that acyclovir in the immunoliposomes was about 10-fold more potent in reducing the virus titer than the free acyclovir. The cytotoxicity of liposomal acyclovir to the uninfected, normal, cells was about 100-fold less than that of the free drug. Therefore, the therapeutic index of acyclovir improved about 1,000-fold by encapsulating it in the target-sensitive immunoliposomes. Similar but more impressive results were also obtained for AraC which showed an increase of 10^7-fold in the therapeutic index in this model system.

Although the results were very encouraging, the immunoliposomes used in these earlier studies were not very stable upon prolonged storage. This is because of the relatively weak bilayer stabilization activity of the acylated antibody. We have recently redesigned the immunoliposomes which showed much improved stability (Pinnaduwage and Huang, 1991). An additional PE bilayer stabilizer, dioleoylphosphatic acid, was included in the lipid composition. There was no content leakage detected with these immunoliposomes when they were stored at 4°C for over 30 days. Furthermore, the immunoliposomes did not destabilize in the absence of the target membrane even if incubated at 75°C for 5 minutes. Although we have not yet tested the activity of these new immunoliposomes for the delivery of proteins and peptides, we expect that they should work well for these applications.

Proteins and Peptides Conjugated to the Long-Circulating Liposomes

One of the important requirements for the efficient target binding of a systemically administered drug is that the drug circulates for sufficiently long periods of time and is not cleared rapidly from the circulation. We have recently shown that long-circulating immunoliposomes can also bind to the target cells with a higher efficiency than the ordinary immunoliposomes which are rapidly cleared from the circulation by the phagocytic cells of the liver and spleen (Maruyama et al, 1990). Most of the proteins and peptides of pharmaceutical interest are rapidly cleared from the circulation rapidly with a $t\frac{1}{2}$ as short as five minutes (Poole et al, 1989). Our strategy to approach this problem is to covalently conjugate the protein or peptide on the surface of the long-circulating liposomes such that the clearance

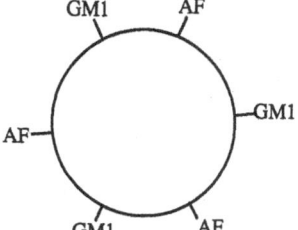

1. AF covalently conjugated to liposome surface.

2. GM1 prolongs circulation of liposomal AF.

3. Liposomal AF still available for function.

Fig. 1. Schematic representation of a liposome design with both long-circulating lipids (GM1) and functional proteins (AF) incorporated in the liposomal membrane.

time of the protein or peptide is prolonged (Fig.1). Asialofetuin (AF) is known for its very short clearance time ($t\frac{1}{2}<1$ min) due to the uptake by the liver hepatocytes via the asialoglycoprotien receptors (Regoeczi et al, 1978). We have covalently conjugated AF to liposomes which contained ganglioside GM_1 to enhance the circulation time (Maruyama et al, 1991). Furthermore, the liposomes were prepared with an average diameter of 200-250 nm which is greater than the average diameter (100-120 nm) of the holes on the fenestrae of the liver sinusoid (Wisse, 1970). Thus, these AF-liposomes could not penetrate the fenestrae to reach the hepatocytes which are primarily responsible for the uptake of free AF. The circulation half-life of AF-liposomes containing GM_1 was up to 150-fold longer than that of the free AF, depending on the AF/lipid ratio of the liposomes (Maruyama et al, 1991). More importantly, the biological activity of the liposomsal AF was indistinguishable to that of the free AF, when assayed on a tissue culture system for the incorporation of free fatty acids into various classes of cellular lipids (Cayatte et al, 1990). This is because the liposomal AF is bound on the surface of the liposomes and is available for interaction with its target cells. Thus, this novel approach of protein/peptide formulation takes advantage of the pharmacokinetic of the long-circulating liposomes, yet has preserved the biological activity of the macromolecule. Similar results have been obtained for the A chain of ricin and the a-factor (a dodecapeptide) of Saccharomyces cerevisiae (Maruyama et al, 1991).

PROTEINS AND PEPTIDES WHICH ACT CYTOPLASMICALLY

Some proteins need to be delivered to the lysosomal compartment of the target cells, for example, antigen molecules to be presented by the MHC class II antigen for the induction of a humoral immune response (Harding et al, 1991). Other proteins have their site of action in the cytosol or nucleus compartment of the target cell. For the former proteins, conventional liposomes ususally do a great job particularly for the delivery of antigen to the macrophages for antigen presentation. However, conventional liposomes do not deliver their contents to the cytosolic compartment with a satisfactory efficiency. We have developed the pH-sensitive liposomes which destabilize at mildly acidic pH (Connor et al, 1984). These liposomes released the contents into the cellular cytosolic compartment more efficiently than the conventional liposomes, probably because they destabilize in the pre-lysosomal compartment of the cell (Collins and Huang, 1989). These liposomes/immunoliposomes have been used for the delivery of dye, drugs, enzymes and DNA (Connor and

Huang, 1985; Collins and Huang, 1987; Zhou et al, 1991). We have recently extended its use to deliver soluble proteins to the cytosolic compartment of the antigen presenting cells (APC) for the MHC class I mediated presentation. Ovalbumin (OVA) could be efficiently entrapped in the pH-sensitive liposomes composed of dioleoyl PE and dioleoyl succinylglycerol (1:1, molar ratio) with repeated freeze-thaw cycles (Zhou et al, 1991). The entrapped OVA was added to mouse thymoma cells (EL4) which were then assayed for cytolysis using OVA specific, cytotoxic T-lymphocytes (CTL). EL4 cells treated with OVA in the pH-sensitive, but not the conventional pH-insensitive, liposomes could be efficiently killed by the CTL (Reddy et al, 1991). Cells treated with free OVA even at 10-fold higher concentration could not be killed. Thus, the pH-sensitive liposomes can deliver soluble proteins and sensitize APC to class I restricted CTL recognition.

More recently we have injected i.v. the liposomal OVA into mice and isolated the splenic dendritic cells from the treated mice. These cells could induce a primary CTL response from a naive T-lymphocyte population which have been co-cultuted with these dendritic cells for 5-7 days (Nair et al, 1991). The CTL was CD8 positive, CD4 negative and OVA-specific. The induction activity was only seen with the pH-sensitive, but not the insensitive liposomes. These preliminary results are very interesting as the induction of a primary CTL response in mouse had not been reported previously. Vaccine development employing the subunits of viral and bacterial components entrapped in the pH-sensitive liposomes will be evaluated in the future.

CONCLUSION

We have discussed here several applications of the novel liposome designs for potential delivery of proteins and peptides. Although the feasibility of the approach has been demonstrated, much remains to be done before an actual pharmaceutical formulation can be realized. For example, the use of the long-circulating liposomes (Ninja Liposomes) (Liu and Huang, 1991) for protein/peptide delivery should be further explored, as the prolonged circulation time presents one of the important hurdles for the pharmaceutical activity of these macromolecular drugs. Our studies only exploited the strategy of the surface conjugated proteins/peptides. Soluble proteins/peptides entrapped in the long-circulating liposomes should also exhibit interesting pharmaceutical activities. This and other aspects of liposomal formulation will undoubtably be studied by various investigators in the near future.

Acknowledgements

Original work done in this lab was supported by NIH grants CA24553 and AI29893. The work on the target-sensitive immunoliposomes and antigen presentation was a collaboration with Dr. Barry T. Rouse of the University of Tennessee, Knoxville.

REFERENCES

Cayatte, A.J., Kumbla, L. and Subbiah, M.T.R., 1990, Marked acceleration of exogenous fatty acid incorporation into cellular triglycerides by fetuin, J. Biol. Chem., 265:5883.
Collins, D., Maxfield, F. and Huang, L., 1989, Immunoliposomes with different acid sensitivities as probes for the cellular endocytic pathway, Biochim. Biophys. Acta, 987:47.

Collins, D. and Huang, L., 1987, cytotoxicity of diphtheria toxin A fragments to toxin-resistant murine cells delivered by pH-sensitive immunoliposomes, Cancer Res., 47:735.

Connor, J., Yatvin, M. and Huang, L., 1984, pH-sensitive liposomes: Acid induced fusion, Proc. Natl. Acad. Sci. U.S.A., 81: 1715.

Connor, J. and Huang, L., 1985, Efficient cytoplasmic delivery of a fluorescent dye by pH-sensitive immunoliposomes, J. Cell Biol., 101:582.

Cullis, P.R. and de Kruijff, B., 1979, Lipid polymorphism and the functional roles of lipids in bioligical membranes, Biochim. Biophys. Acta, 559:399.

Ellens, H., Bentiz, J. and Szoka, F., 1986, Fusion of phosphatidyl-ethalonamine-containing liposomes and mechanism of the L_α-H_{II} phase transition, Biochemistry, 25:41.

Fishman, Y. and Citri, N., 1975, L-asparaginese entrapped in liposomes: preparation and properties, FEBS Lett., 60:17.

Harding, C.V., Collins, D.C., Slot, J.W., Geuze, H.J. and Unanue, E.R., 1991, Liposome-encapsulated antigens are processed in lysosomes, recycled, and presented to T cells, Cell, 64:393.

Ho, R.J.Y., Rouse, B.T. and Huang L., 1986a, Target-sensitive immuno-liposomes: preparation and characterization, Biochemistry, 25:5500.

Ho, R.J.Y., Rouse, B.T. and Huang, L., 1986b, Destabilization of phos-phatidyl-ethanolamine immunoliposomes by antigen binding: A valuable approach for rapid virus detection, Biochem. Biophys. Res. Comm., 138:931.

Ho, R.J.Y., Rouse, B.T. and Huang, L., 1987a, Target-sensitive immuno-liposomes as an efficient drug carrier for antiviral activity, J. Biol. Chem., 262:13973.

Ho, R.J.Y., Rouse, B.T. and Huang, L., 1987b, Interactions of target sensitive immunoliposomes with herpes simplex virus: The foundation of a sensitive immunoliposome assay for the virus, J. Biol. Chem., 262:13979.

Liu, D. and Huang, L., 1991, Immunoliposome targeting to pulmonary endothelium, in: "Drug Delivery Series, Vol. 2", Florence, A.T. and Gregoriadis, G., eds., in press.

Manning, M.C., Patel, K. and Borchardt, R.T., 1989, Stability of protein pharmaceuticals, Pharm. Res., 6:903

Maruyama, K., Kennel, S. and Huang, L., 1990, Lipid composition is important for highly efficient target binding and retention of immunoliposomes, Proc. Natl. Acad. Sci. U.S.A., 87:5744

Maruyama, K., Mori, A., Bhadra, S., Subbiah, M.T.R., and Huang, L., 1991, Protein and peptides bound to long-circulating liposomes, Biochim. Biophys. Acta, in press.

Nair, S. Zhou, F., Reddy, R., Huang, L. and Rouse, B.T., 1991, Soluble proteins delivered to dendritic cells via pH-sensitive liposomes induce primary CTL response in vitro, J. Exp. Med., 175:609.

Pinnaduwage, P. and Huang, L., 1991, Stable target-sensitive immuno-liposomes, in preparation.

Poole, S., Tymkewycz, P.M., Bird, T. and Saklatvala, J., 1989, Bioavailability and tissue-targeting of therapeutic proteins, in: "Therapeutic Peptides and Proteins: Formulation, Delivery and Targeting", Marshak, D. and Liu, D., eds., Cold Spring Harbor Lab., New York.

Reddy, R., Zhou, F., Huang, L. and Rouse, B.T., 1991, pH-sensitive liposomes provide an efficient means of sensitizing target cells to class I restricted CTL recognition of a soluble protein, J. Immunol. Method, 141:157.

Regoeczi, E., Dabanne, M.T., Hatton, M.W.C. and Koj, A., 1978, Elimination of asialofetuin and asialorosomucoid by the intact rat, Biochim. Biophys. Acta, 541:372.

Wisse, E., 1970, An electronic microscopic study of the fenestrated endo-
    thelial lining of rat liver sinusoids, J. Ultastructure Res.,
    31:125.
Zhou, F., Rouse, B.T. and Huang, L., 1991, An improved method of loading
    pH-sensitive liposomes with soluble proteins for class I restricted
    antigen presentation, J. Immunol. Method, 145:143.
Zhou, X., Klibanov, A. and Huang, L., 1991, Improved encapsulation of DNA
    in pH-sensitive liposomes for transfection, J. Liposome Res., in
    press.

STERICALLY STABILIZED LIPOSOMES AS DRUG CARRIERS: PHARMACOKINETICS,

TISSUE DISTRIBUTION AND THERAPEUTIC EFFECTS IN TUMOUR-BEARING MICE

A. Gabizon[1], S.K. Huang[2], and D. Papahadjopoulos[2]

[1]Sharett Institute of Oncology, Hadassah University
Hospital, Jerusalem, Israel
[2]Cancer Research Institute, University of California
San Francisco, California, USA

INTRODUCTION

Recent advances in liposome formulation have resulted in a new generation of phospholipid vesicles with improved pharmacokinetic properties (Gabizon and Papahadjopoulos, 1988; Allen et al, 1989; Klibanov et al, 1990; Blume and Cevc, 1990; Senior et al, 1991; Papahadjopoulos et al, 1991). These liposomes also referred to as Stealth[R] (Allen, 1989) circulate for prolonged periods of time with stable retention of their contents. They appear to evade recognition by the reticulo-endothelial system (RES), as indicated by their delayed and diminished accumulation in liver and spleen as compared with conventional liposome formulations. They also show a significant localization in transplantable mouse and human tumours (Gabizon et al, 1990), a phenomenon apparently related to the increased microvascular permeability of tumours.

The critical factors required to obtain these long-circulating liposomes are (Gabizon and Papahadjopoulos, 1988; Lasic et al, 1990; Gabizon and Papahadjopoulos, 1992): (a) A small vesicle size (100 nm diameter or less). The effect of size on liposome circulation time has been well documented (Senior, 1987). It has specifically a great impact on spleen liposome uptake (Gabizon and Papahadjopoulos, 1992). (b) Use of phospholipids of high phase transition temperature (T_c > 37°C) and cholesterol as the main vesicle components. This issue is related mainly to vesicle stability, i.e. the ability of liposomes to retain their contents for prolonged circulation times. Both, high T_c phospholipids and cholesterol, have been known as stabilizing factors when liposomes are exposed to plasma in vitro or in vivo at physiologic temperature (Gregoriadis, 1988). (c) Presence of surface hydrated groups anchored on a negatively-charged phospholipid. This has been the major recent contribution leading to the design of Stealth liposomes. The presence of surface hydrated groups, as in the case of polyethylene-glycol derivatized distearoyl phosphatidylethanolamine (PEG/DSPE), appears to be required to achieve a high degree of steric stabilization. this effect may be related to an inhibition of the binding of opsonins to the liposome surface and other hydrophobic interactions conducive to vesicle aggregation (Lasic et al, 1991). Traditionally, negative charge has been considered as a factor enhancing Kupffer cell liposome uptake and

Targeting of Drugs 3: The Challenge of Peptides and Proteins
Edited by G. Gregoriadis et al., Plenum Press, New York, 1992

51

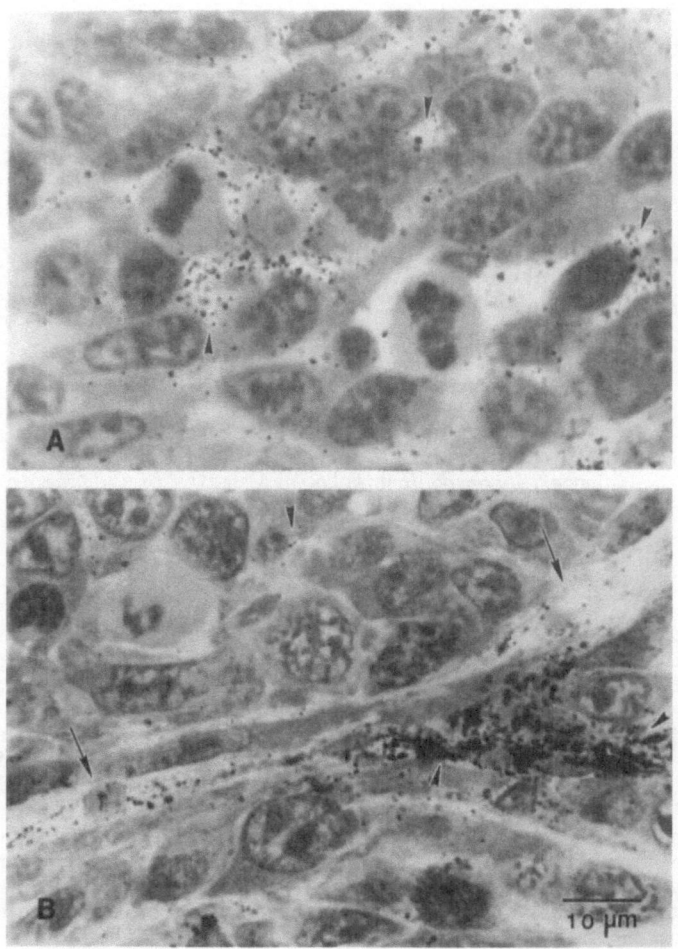

Fig. 1. Thick section of C-26 colon carcinoma implanted
        subcutaneously (A) and directly into the liver (B) of
        BALB/c mice. Coloidal gold containing liposomes composed
        of PEG/DSPE-EPC-Chol (molar ratio, 0.8-10-5) were injected
        IV into mice. Tissues were collected and fixed 24 hr
        later. In A, the silver-enchanced gold particles (arrow
        heads) are seen in the extracellular space of the tumor.
        In B, the dense, silver-enhanced particles (arrow heads)
        are observed within the penetrating vessels filled with
        erythrocytes (arrows), as well as in the surrounding tissue
        of the tumor.

detrimental to liposome circulation time. However, in recent years, it
has become clear that a small fraction of negatively-charged lipid in the
form of PEG/DSPE, monosialo-ganglioside (GM1), or phosphatidylinositol
(PI) can improve liposome circulation time. One key characteristic of
these negatively-charged lipids is that the negatively charged head group
is sterically shielded by a bulky hydrophilic moiety. In this way, the

negative charge appears to contribute to the steric stabilization of liposomes while the interaction with circulating or cell surface opsonins is prevented (Gabizon and Papahadjopoulos, 1992).

Clearly, one of the potential applications for sterically stabilized liposomes is the delivery of anticancer agents to tumours. We will present here morphological observations confirming the localization of these liposomes in experimental tumours and results of preclinical pharmacological studies using a hydrogenated PI (HPI)-containing formulation as carrier of doxorubicin (Gabizon et al, 1989; Gabizon, 1992).

METHODOLOGY

All the methodology and sources of materials involved in the experiments presented here have been described in detail elsewhere. The technique used for the preparation and morphological studies with silver-enhanced colloidal gold-labelled liposomes has been reported by Huang et al (1991). Regarding the biodistribution, toxicity, and anti-tumour studies with liposome-encapsulated doxorubicin (DOX) and free DOX, the sources of materials and methods used, including liposome preparation, drug encapsulation and drug determination in tissues have been extensively described in previous publications (Gabizon et al, 1989; Gabizon, 1992).

RESULTS AND DISCUSSION

Localization of colloidal gold containing liposomes in a mouse tumour

In these experiments, we used liposomes composed of PEG/DSPE, egg phosphatidylcholine (EPC), and cholesterol (Chol). There was no need to use a high T_c phospholipid such as hydrogenated PC (HPC), since there is no problem of leakage with colloidal gold. Localization of colloidal gold-containing liposomes in mouse colon carcinoma (C-26) implanted tumour was studied by light microscopy of silver-enhanced thick sections (Fig. 1, A and B). In some regions of the tumour, silver-enhanced colloidal gold particles are found predominantly in the extracellular space between tumour cells. High density of silver-enhanced colloidal gold can often be seen in the area surrounding blood vessels which penetrate into the tumour mass. In some areas, it was clear that gold particles, presumably encapsulated in liposomes, can cross the blood vessel endothelium, extensively penetrating into the extravascular/interstitial space between tumour cells. In other regions, gold particles could be seen to accumulate in small vessels without any penetration into the surrounding tissue. Similar preparations of colloidal gold particles, when injected without liposomes, were not found to accumulate in tumours.

Biodistribution studies

As seen in Fig. 2, DOX encapsulated in HPI-HPC-Chol liposomes, referred heretofore as HPI-DOX, is cleared very slowly from plasma. Most of the liposome-associated drug is cleared with a distribution half-life of about 15 hours. Significant concentrations of liposome-associated DOX are detected as long as 3 days after injection. The values for apparent volume of distribution and clearance of liposome-associated drug are about 200-fold smaller than those of free drug. Thus, the use of sterically stabilized liposomes as DOX carriers causes a drastic change in the pharmacokinetic parameters of the encapsulated drug.

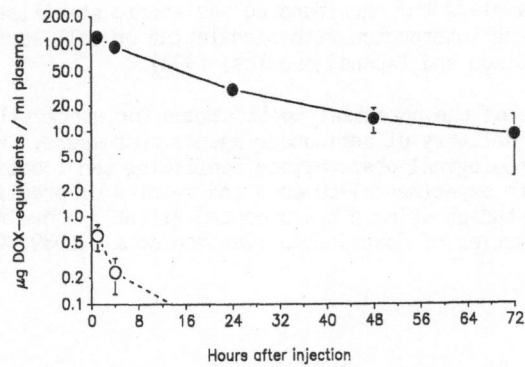

Fig. 2. Plasma drug levels after injection of free DOX (--o--) or
HPI-DOX (—o—). BALB/c mice received IV a dose of
10 mg/kg DOX in free or liposome-encapsulated form (HPI-
HPC-Chol, molar ratio 1-10-8).

Figures 3 and 4 show the results of DOX levels in liver and heart of
mice injected with free DOX or HPI-DOX. The peak levels in liver (Fig. 3)
were similar for both forms of treatment. However, while the peak level
for free DOX was attained as soon as 30 min after injection, the peak level
for HPI-DOX was reached only 24 hr after injection. The delayed accumulation of the latter is in agreement with its slow plasma clearance rate
shown above. Regarding the heart muscle (Fig. 4), there was a significant
reduction in drug concentration using DOX encapsulated in HPI liposomes.
Most importantly, the peak concentration seen shortly after injection of
free DOX, and believed to play a central role in the cardiotoxic effect,
was completely avoided when HPI-DOX was administered. Thus, encapsulation
of DOX in Stealth liposomes is likely to reduce anthracycline-induced
cardiotoxicity, enabling perhaps safer administration of greater cumulative
doses of the drug.

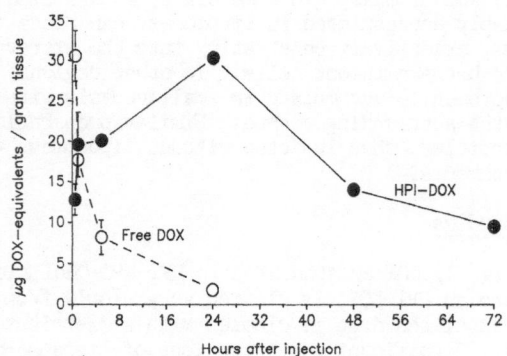

Fig. 3. Drug levels in the liver after injection of free DOX
(--o--) or HPI-DOX (—●—). BALB/c mice received IV a
dose of 10 mg/kg DOX in free or liposomes-encapsulated form
(HPI-HPC-Chol, molar ratio 1-10-8).

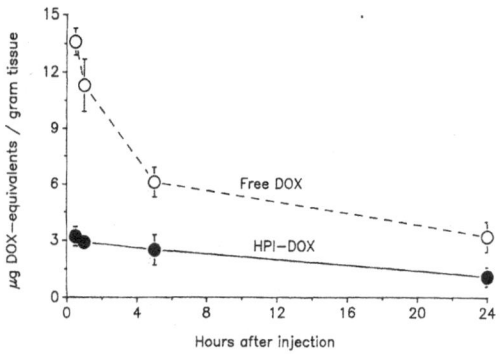

Fig. 4. Drug levels in the heart after injection of free DOX
(--o--) or HPI-DOX (—•—). BALB/c mice received IV a dose
of 10 mg/kg DOX in free or liposome-encapsulated form (HPI-
HPC-Chol, molar ratio 1-10-8).

Since sterically stabilized liposomes localize in tumours in
substantial amounts (Gabizon et al, 1990), one would expect a significant
enhancement in the concentration of a liposome-encapsulated drug such as
DOX, as compared to conventional free drug administration. This was
investigated using a solid implant of the J6456 lymphoma in the mouse
thigh. The results shown in Fig. 5 indicate that the drug levels
obtained after HPI-DOX injection are about 4-fold higher than those after
free DOX injection. In addition, the peak tumour drug level with HPI-DOX
occurs late after drug injection, stressing the fact that liposome
accumulation in tumours is a slow process requiring a long circulation
time. In contrast to tumour, there was no difference between HPI-DOX and
free DOX regarding drug levels in a normal tissue, such as muscle (see
also Fig. 5).

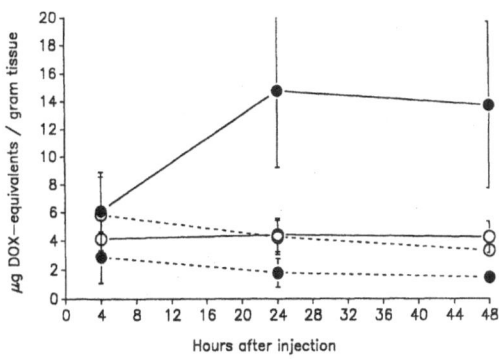

Fig. 5. Drug levels in tumor and muscle after injection of free DOX
(tumor, —o—; muscle, --o--) or HPI-DOX (tumor, —•—;
muscle, --•--). BALB/c mice bearing 300-700 mg tumor
implants of the J6456 lymphoma in the left hind leg
received IV a dose of 10 mg/kg DOX in free or liposome-
encapsulated form (HPI-HPC-Chol, molar ratio 1-10-8).

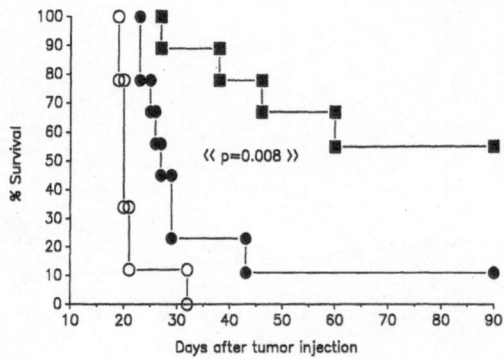

Fig. 6. Lethal effect of escalated doses of free DOX (--o--) or
HPI-DOX (--●--). BALB/c mice received IV DOX in either
free or liposome-encapsulated form (HPI-HPC-Chol, 1-9-8)
and were followed up for 90 days.

Toxicity studies

Fig. 6 shows the lethal effect of escalating doses of free and HPI-
DOX in mice. The lethal events are the result of acute toxicity within 2
weeks after injection, and delayed toxicity which takes place
approximately 2 months after injection. The former is generally related
to gastrointestinal toxicity, while the latter is the combined result of
nephrotoxicity and cardiotoxicity. Clearly, the toxicity of DOX in HPI
liposomes is significantly reduced as compared to that of free DOX in
mice. While the LD_{50} of free DOX is between 10 to 15 mg/kg, that of HPI-
DOX is between 20 to 25 mg/kg.

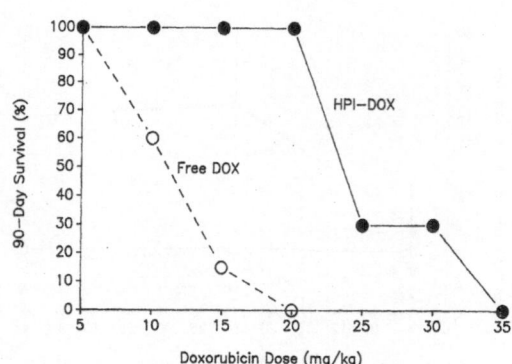

Fig. 7. Therapeutic effect of free DOX and HPI-DOX in the ascitic
J6456 tumor model. BALB/c mice received an IP inoculum of
10^6 J6456 lymphoma cells and were treated IV, 5 days later,
with DOX (10 mg/kg) in either free or liposome-encapsulated
form (HPI-HPC-Chol, molar ratio 1-9-8). Untreated, o; Free
DOX treated, ●; HPI-DOX treated, ■.

Therapeutic Efficacy

Obviously, the key study to evaluate HPI-DOX is to compare its therapeutic efficacy with that of free DOX. For that, we chose an ascitic form of the J6456 lymphoma. Treatment was administered IV, 5 days after IP tumour inoculation. As seen in Fig. 7, median survival and cure rate were significantly increased using HPI-DOX. The dose given was subtoxic (10 mg/kg) and the same for both forms of treatment. Therefore, this observation was not due to a difference in toxicity, but to a net gain in antitumour activity using the liposome-encapsulated formulation.

In conclusion, small-sized liposomes which are sterically stabilized with the help of surface hydrated groups and a small amount of negative charge can circulate for long periods of time due to reduced uptake by the RES. A significant fraction of these liposomes can localize in implanted tumours. Such liposomes have been used as carriers of a cytotoxic drug, doxorubicin, resulting in: a) improved drug delivery to the tumour; b) protection of a doxorubicin-sensitive organ, the heart; c) reduction of systemic toxicity; and, d) enhanced anti-tumour therapeutic efficacy. These observations are encouraging, adding momentum to the application of liposomes in cancer therapy.

Acknowledgements

This work was supported by a grant from the Israel Cancer Research Fund to A.G., NIH grant CA-35349 to D.P., and grants from Liposome Technology Inc. to A.C. and S.K.H.

REFERENCES

Allen, T. M., 1989, Stealth™ liposomes: avoiding reticuloendothelial uptake, in: "Liposomes in the therapy of infectious diseases and cancer", G. Lopez-Berestein and I.J. Fidler (eds.), Alan R. Liss Inc., New York.

Allen, T. M., Hansen, C. and Rutledge, J., 1989, Liposomes with prolonged circulation times: factors affecting uptake by reticuloendothelial system and other tissues, Biochim. Biophys, Acta, 981:27.

Blume, G. and Cevc, G., 1990, Liposomes for sustained drug release in vivo, Biochim. Biophys. Acta, 1029:91.

Gabizon, A., 1992, Selective tumor localization and improved therapeutic index of anthracyclines encapsulated in long-circulating liposomes, Cancer Res., 52:891.

Gabizon, A. and Papahadjopoulos, D., 1988, Liposome formulations with prolonged circulation time in blood and enhanced uptake by tumors, Proc. Natl. Acad, Sci. USA, 85:6949.

Gabizon, A. and Papahadjopoulos, D., 1992, The role of surface charge and hydrophilic groups on liposome clearance in vivo, Biochim. Biophys. Acta, 1103:94.

Gabizon, A., Shiota, R. and Papahadjopoulos, 1989, Pharmacokinetics and tissue distribution of doxorubicin encapsulated in stable liposomes with long circulation times, J. Natl. Cancer Inst., 81:1484.

Gabizon, A., Price, D. C., Huberty, J., Bresalier, R. S. and Papahadjopoulos, D., 1990, Effect of liposome composition and other factors on the targeting of liposomes to experimental tumors: biodistribution and imaging studies, Cancer Res., 50:6371.

Gregoriadis, G., 1988, Fate of injected liposomes: observations on entrapped solute retention, vesicle clearance and tissue distribution, in: "Liposomes as Drug Carriers: Recent Trends and Progress", G. Gregoriadis (ed.), John Wiley and Sons, Chichester.

Huang, S. K., Lee, K. D., Woodle, M. C., Lasic, D. D., Redemann, C., 1991, Light microscopic localization of silver-enhanced liposome-entrapped colloidal gold in mouse tissues, Biochim. Biophys, Acta, 1069:117.

Klibanov, A. L., Maruyama, K., Torchilin, V. P. and Huang, L., 1990, Amphipathic polyethyleneglycols effectively prolong the circulation time of liposomes, FEBS Lett., 268:235.

Lasic, D. D., Martin, F. J., Gabizon, A., Huang, S. K. and Papahadjopoulos, D., 1991, Sterically stabilized liposomes: a hypothesis on the molecular origin of the extended circulation times, Biochim. Biophys. Acta, 1070:187-192.

Papahadjopoulos, D. Allen, T. M., Gabizon, A., Mayhew, E., Matthay, K., Huang, S. I., Lee, K.D., Woodle, M.C., Lasic, D. D., Redemann, C. and Martin, F. J., 1991, Sterically stabilized liposomes: improvements in pharmacokinetics and antitumor therapeutic efficacy, Proc. Natl. Acad. Sci. USA, 88:11460.

Senior, J. H., 1987, Fate and behaviour of liposomes in vivo: a review of controlling factors, CRC Crit. Rev. Ther. Drug Carrier Syst., 3:123.

Senior, J., Delgado, C., Fisher, D., Tilcock, C. and Gregoriadis, G., 1991, Influence of surface hydrophilicity of liposomes on their interaction with plasma protein and clearance from the circulation: studies with poly(ethyleneglycol)-coated vesicles, Biochim. Biophys. Acta, 1062:77.

LIPOSOMES AS IMMUNOLOGICAL ADJUVANTS

Gregory Gregoriadis

Centre for Drug Delivery Research, The School of Pharmacy
University of London, 29-39 Brunswick Square, London
WC1N 1AX, UK

INTRODUCTION

Most vaccines already in use have evolved from attenuated or killed
organisms and, as a result, they may produce unacceptable side effects. On
the other hand, subunit vaccines produced from biological fluids, may be
contaminated with known or unknown infectious agents. The emergence of
recombinant DNA and monoclonal antibody technology and recent rapid progress
in the understanding of the immunological structure of proteins and of
factors regulating immune responses have laid the foundations of a new
generation of recombinant subunit and synthetic peptide vaccines represent-
ing small regions of microbial proteins. Such vaccines are capable of
inducing specific immune responses that control infectious micro-organisms
and, as they are defined at the molecular level, they are potentially safe.
However, subunit and peptide vaccines are only weakly or non-immunogenic.

A wide variety of structurally unrelated agents (immunological
adjuvants) can promote immune responses to weak antigens (Bomford, 1985).
Adjuvants include aluminum hydroxide (alum), saponins, Pluronic block
polymers in mineral oil, killed mycobacteria in mineral oil (Freund's
complete adjuvant; FCA), bacterial products (eg. lipopolysaccharide (LPS)
and muramyl dipeptide (MDP)) and liposomes (for reviews on a number of
adjuvants see Gregoriadis et al, 1989). It is believed (Allison and Byars,
1986) that immunological adjuvants produce their effect by the creation of a
depot (for instance, alum, oil emulsions) at the area of injection. Release
of antigens and their interaction with antigen presenting cells (APC) is
thus prolonged. APC also invade the depot area because of local inflam-
mation, and adjuvant-antigen complexes may migrate to the regional lymph
nodes containing T-cells. Alternatively or simultaneously, there may be
activation of macrophages (for instance by LPS or MDP) which then release
interleukin-1 (IL-1). The concerted action of IL-1 and antigen on T-cells
produces interleukin-2 (IL-2) as well as other mediators which activate
effector T cells to produce cell-mediated immunity (CMI) or stimulate the
production of antibody forming B cells (humoural immunity; HI). It is
thought (Allison and Byars, 1986) that strong and persistent immune
responses may be elicited by directing antigens to interdigitating cells
which constitutively express class II major histo-compatibility antigens and
to follicular dendritic cells. Both types of cells specialise in presenting antigen to T and B lymphocytes respectively, and may thus favour
selectively the induction of CMI or HI.

Targeting of Drugs 3: The Challenge of Peptides and Proteins
Edited by G. Gregoriadis et al., Plenum Press, New York, 1992

Most adjuvants currently available are toxic and are known to induce granulomas, acute and chronic inflammations (FCA), cytolysis (saponins, some Pluronic polymers) and pyrogenicity, arthritis and anterior uveitis (LPS and MDP). It is however possible to curtail the toxicity of some of the adjuvants (without major loss of their adjuvanticity) by appropriate changes in the agent's structure (LPS, MDP) or by the judicious choice and synthesis of adjuvant analogs (saponin, Pluronic polymers) (Allison and Byars, 1986). Nevertheless, alum remains, for well over half a century, the only adjuvant licenced for use in humans. Alum, however, is not always effective and increases CMI only slightly if at all. Thus, efforts are being made (Gregoriadis et al, 1989; Allison and Byars, 1986) towards the development of safe and effective adjuvants to meet the challenges of new-generation vaccines.

New adjuvants should comply with a number of criteria which include certain practical ones, for example inexpensive raw materials, simple manufacture, stability under storage even at elevated temperatures, and availability in a freeze-dried form when required. Modern adjuvants are also expected to be biodegradable, non-toxic and non-immunogenic. Furthermore they should, ideally, be capable of inducing both CMI and HI to antigens administered by a variety of routes and also act synergistically with other adjuvants. Most of these criteria are satisfied by liposomes which are best known for their potential and actual uses in targeted drug delivery (Gregoriadis, 1988a). Here I shall discuss liposomes as immunological adjuvants and will begin with comments on relevant technology as applied in the laboratory and on the behaviour of liposomes in vivo. The latter is highly relevant to understanding the immunoadjuvant activity of the system and to attempts to improve such activity.

LIPOSOME TECHNOLOGY

Manufacture of liposomal antigen products for clinical or veterinary use requires methodology for efficient antigen incorporation in vesicles of a narrow size distribution using simple, reproducible and inert (to antigens) technology (Gregoriadis, 1984; Lopez-Berestein and Fidler, 1989). During the 80's, considerable progress was made in this respect and a number of well defined formulations incorporating active agents including antigens can now be produced in a stable form mainly by liposome biotechnology companies but also by other industries. Some of these formulations are being tested in clinical trials (Lopez-Berestein and Fidler, 1989) and a few (eg. a liposomal amphotericin-B product and a liposome-based veterinary vaccine), are already licenced. However, most of the techniques developed, have limitations. For instance, they may be applicable only to certain drugs of low molecular weight or require the use of detergents, sonication or organic solvents in the presence of the drug destined for entrapment (Gregoriadis, 1984; Lopez-Berestein and Fidler, 1989). No such conditions are present in a simple and amenable to scale-up procedure developed in the author's laboratory (Kirby and Gregoriadis, 1984). The technique produces multilammellar vesicles by a process based on the dehydration of buffer or water-loaded SUV in the presence of free drugs. Subsequent rehydration leads to high drug entrapment yields. Thus, entrapment values obtained for a variety of antigens and immunomodulators (tetanus toxoid (Gregoriadis et al, 1987), influenza virus subunit proteins (Tan et al, 1989), recombinant hepatitis B surface antigen (Gregoriadis et al, 1989), Leishmania major antigens (Kahl et al, 1989), poliovirus 3-VP2 peptides (Xiao et al, 1989) and recombinant IL-2 (Tan and Gregoriadis, 1989b) (DRV) were high and reproducible for each of the proteins. In addition, freeze-drying of DRV in the presence of a cryoprotectant leads to the retention of most of the antigen content within intact vesicles formed on reconstitution with saline (Gregoriadis et al, 1987).

It has also been shown (Gregoriadis et al, 1990) that microfluidization of solute-containing DRV composed of equimolar phospholipid (up to 66 μmoles) and cholesterol produces smaller vesicles (mean diameter of about 100 nm) which retain much of the originally entrapped solute. Retention of solutes was found (Gregoriadis, 1990) to depend on the number of microfluidization cycles, the medium (water or phosphate-buffered saline) in which microfluidization was carried out and on whether or not unentrapped solute (after DRV formation) was present during microfluidization. Under appropriate conditions, liposomes with mean diameters of 200 nm or less were generated, containing about 35-78% of the originally entrapped solute. This approach compares favourably with other procedures (Mayhew et al, 1984) which use larger amounts of lipids (180 to 300 μmols) to achieve similarly efficient entrapment and provides formulations with augmented solute to lipid mass ratios and, therefore, more economical and, potentially, less toxic.

BEHAVIOUR OF LIPOSOMES IN VIVO

Liposomes were first proposed and tested in animals as a drug carrier in 1970 (for a historical background see Gregoriadis, 1976). Since then, much of their behaviour in vivo and ways to control it, have become known (Gregoriadis, 1988b). Relevant to most applications (including immunoadjuvant activity in vaccines) is, in this respect, the effect of biological fluids (with which liposomes come into contact on injection) on (a) the liposomal structural integrity, and (b) the rate at which vesicles are cleared from the site of injection and distributed in tissues. Liposomal behaviour in all these events, is determined by the structural characteristics of the vesicles. It is thus known (Gregoriadis, 1988b) that plasma high density lipoproteins (HDL) remove phospholipid molecules from intravenously injected conventional liposomes (eg. those made of egg phosphatidylcholine; PC). Liposomes attacked by HDL are known to become very leaky or to disintegrate and release their drug contents. However, by substituting PC with "high melting" phospholipids such as distearoyl phosphatidylcholine (DSPC) or sphingomyelin, or by supplementing phospholipids with excess cholesterol, vesicle bilayers become rigid at 37°C (the body's temperature) or have their phospholipid molecules packed. They thus resist phospholipid loss to HDL, their integrity is preserved and entrapped solutes (eg. antigens) remain with the vesicles in the presence of blood for longer periods of time.

Early observations (Gregoriadis, 1988b) revealed that liposomal stability (in terms of solute retention) in vivo is directly related to the vesicle rate of clearance from the blood circulation (ie. the more stable the vesicles are the longer they circulate in the blood). Such a relationship between vesicle clearance and vesicle stability is abolished when a negative or a positive surface charge is present on the bilayer surface (Tan and Gregoriadis, 1989a; Senior et al, 1985). For example, even the most stable, negatively charged liposomes exhibit short half-lives (Senior et al, 1985). Reduction in the half-life of stable liposomes also occurs as their size increases (Senior et al, 1985). Several groups (Blume and Cevc, 1990; Senior et al, 1991; Papahadjopoulos et al, 1992) have shown recently that circulation time of liposomes can be prolonged significantly by rendering their bilayer surface highly hydrophilic, for instance, by the use of polyethyleneglycol covalently coupled to a liposomal phospholipid component. Perhaps not surprisingly, liposomes, which exhibit long half-lives either because of their lipid composition or because of a highly hydrophilic surface are taken up by the reticulo-endothelial system (RES) much more slowly. As expected (Gregoriadis, 1976), such long-lived liposomes are especially amenable to targeting to cells with which they do not normally interact in vivo. Targeting is usually promoted by the coupling of cell-specific ligands (eg. antibodies and certain glycoproteins and glycolipids) onto the liposomal surface (Gregoriadis, 1988a). Regardless of the mechanism of interaction of liposomes with cells, their uptake (when it occurs) proceeds through endocytosis, although fusion may be involved to some extent (Gregoriadis, 1988a). Work by Huang and Zhou (this book) has also shown that the lysosomotropic pathway of liposomal entry into cells can be interfered with in vitro by the use of liposomes which, owing to their lipid composition, can fuse with the endocytic vacuoles and release their contents into the cytoplasm before the vacuoles can fuse with the lysosomes. It therefore appears possible to deliver antigens and drugs into more than one of the cell's compartments.

Knowledge of the fate of liposomes injected intravenously is also relevant to preparations administered by alternative parenteral routes. For example, a proportion of liposomes which is determined by their size, lipid composition and route of injection, enters the lymphatic and, eventually, the blood circulation where they behave as if given intravenously (Gregoriadis, 1988a; Turner et al, 1983). However, whereas liver, spleen and bone marrow intercept much of the dose of liposomes given by the intravenous route, these tissues will account only for a small fraction of the dose administered by other routes. The remainder (eg. for liposomes injected subcutaneously or intramuscularly) is either retained at the site of injection (Turner et al, 1983) or ends up in the lymph nodes draining the injected site. Indeed, uptake by lymph nodes is much greater (over 100-fold in terms of percent uptake per gram tissue) than with the other RES tissues (Gregoriadis, 1988a; Turner et al, 1983).

Attempts by several workers (Gregoriadis, 1988a) to establish whether liposomes given by the intragastric route promote the absorption of drugs and proteins have not given conclusive results. Although there is evidence (Gregoriadis, 1988a) that drugs such as tubocurarine, insulin, factor VIII, anticoagulants and vitamins reach the blood circulation when given in the liposome form, their absorption is only minimal and also unpredictable. Nonetheless, liposomes which are resistant to bile salts or phospholipase attack (eg. those made of DSPC) protect agents from the hostile gut milieu. Some of these liposomes survive for long enough to enter the lymphatics and this would explain the immunoadjuvant action of the system in animals immunised orally (Gregoriadis, 1988a).

LIPOSOMES AS IMMUNOLOGICAL ADJUVANTS

Early studies (Gregoriadis, 1976) with a liposome-entrapped protein (diphtheria toxoid) revealed that liposomes acted as an immunological adjuvant augmenting antibody response to the protein in mice to levels higher than those observed with the free protein (Allison and Gregoriadis, 1974). The adjuvanticity of liposomes was subsequently confirmed and extended to include a large variety of antigens (Gregoriadis, 1990) derived from bacteria, protozoa, viruses, tumours, spermatozoa etc. Moreover, elicited immunity was protective in several of the animal models (Gregoriadis, 1990). Significantly, none of the studies reported any liposome-induced side reactions such as granulomas, etc.

It has been shown that physical association between liposomes and antigen (as opposed to their presence in a mixture independently of each other) is a prerequisite for immunoadjuvant action to occur (Shek, 1984). Physical association can be in the form of antigen entrapped within the aqueous phase of the vesicles, electrostatic adsorption of antigen onto the bilayer surface or, for membrane-soluble antigens, hydrophobic insertion into the lipid phase. However, with the exception of certain viral subunits that are known to form well organized structures in conjunction with lipo-

somal phospholipids (virosomes) (Almeida et al, 1975) accommodation of
antigens within the bilayers is largely undefined for most liposomal antigen
formulations studied so far (Gregoriadis, 1990). Similarly undefined are
other liposomal characteristics such as vesicle size distribution, number of
lamellae within vesicles, etc. Judging from the multitude of different
lipid compositions already investigated (Gregoriadis, 1990), it appears that
liposomal adjuvanticity does not depend on any specific formulation nor does
it seem to be especially related to the mode of immunization: the choice of
antigen and liposomal lipid dose and route and frequency of injections has
often been arbitrary (Gregoriadis, 1990).

Although the design of liposomal antigen formulations and immunization
protocols has been largely empirical (Gregoriadis, 1990), adjuvanticity of
liposomes was observed in terms of primary and/or secondary responses
against all antigens tested for HI (Gregoriadis, 1990). Moreover, T cell
dependency of antigens was demonstrated (Shek, 1984) in at least one case
studied. With regard to HI, recent work (Davis et al, 1987) has shown that
liposomal adjuvanticity occurs during primary immunization and is observed
with most IgG subclasses. Further, there was no apparent shift in subclass
responses compared to responses seen with the free antigen (Almeida et al,
1975). Information on liposomal fate in vivo as already outlined, suggests
that antibody production is stimulated partly or wholly as a result of the
system's function as an antigen depot. Indeed, liposomes are likely to
supply macrophages with released and/or entrapped antigen at rates which
favour its efficient processing by the cells and eventual presentation.
Macrophage involvement in liposomal adjuvanticity (a strong possibility in
view of T cell participation) (Shek, 1984) is, in fact, supported by studies
showing its absence in animals depleted of their macrophages (Shek, 1984; Su
and van Nieuwmegen (1989).

Induction of CMI by liposomes is probably one of their most important
features as an immunoadjuvant. Evidence to support CMI has been provided by
positive DTH reactions (Manesis et al, 1979), the lymph node lymphocyte
proliferative response test and the induction of cytotoxic T lymphocytes
(for further details see Shek, 1984). Liposome-induced CMI is unlikely to
be the result of the antigen-depot mechanism since adjuvants such as oil
emulsions and alum acting in this way, induce only or predominantly HI. A
more plausible explanation of liposome-induced CMI could be related to the
presentation of antigen in a hydrophobic microenvironment. For instance, it
is known that proteins conjugated to lipids induce DTH in proportion to the
lipid's hydrophobicity (Dailey and Hunter, 1977). This, in turn, increases
uptake of the complex by macrophages thus improving antigen presentation to
T cells. Such events may also be favoured by the efficient (and selective)
localization of liposomal antigens into the regional lymph nodes (Tumer et
al, 1983).

AMPLIFICATION OF LIPOSOMAL ADJUVANTICITY

Amplification of liposomal immunoadjuvant action has been achieved
through receptor-mediated targeting to macrophages (Garcon et al, 1988), the
use of co-adjuvants or modification of the vesicles' structural characteristics (Gregoriadis, 1990). Immune responses to liposomal antigens are, for
instance, further augmented through their administration together with other
adjuvants (for examples see Gregoriadis, 1990). Co-adjuvants can be
entrapped together with the antigen in the same vesicles, entrapped in
separate vesicle populations which are mixed before injection (eg. IL-2,
lipid A) or, simply, mixed with antigen-containing liposomes (eg. alum, B.
pertussis). It has been shown for instance (Tan and Gregoriadis, 1989b)
that immune responses are increased above or reduced below levels achieved
with the liposomal antigen alone, depending on the amount of cytokine (IL-2)

given, the mode of presentation with the liposomal antigen (co-entrapped or separately entrapped) and the IgG subclass tested.

Previous comments on related mechanisms for liposome-mediated HI and CMI suggest that liposomal adjuvanticity reflects the vesicular structure of the system and, probably, its lipid nature rather than being the result of lipid composition or other liposomal characteristics (eg. vesicle size, lamellarity, surface charges, etc). Such characteristics, however, effect-ively control liposomal behaviour in vivo (Gregoriadis, 1988a) and are thus likely to be instrumental in the way immunoadjuvant activity is expressed, both qualitatively and quantitatively. The extent to which bilayer fluidity (Davis and Gregoriadis, 1987; Bakouche and Gerlier, 1986; Kinsky, 1978) number of lammellae in bilayers (Shek, 1984), vesicle size (Francis et al, 1985) and surface charge (Davis and Gregoriadis, 1987; Kraaijeveld et al, 1984; Latif and Bachhawat, 1987), lipid to antigen mass ratio (Davis and Gregoriadis, 1987; Davis and Gregoriadis, 1989) and mode of antigen local-ization (Davis and Gregoriadis, 1987; Shek, 1984; van Rooijen and van Nieuwmegen, 1980) within liposomes influence, one way or another, adjuvant-icity have been therefore studied extensively. They were all shown to have an effect but conclusions as to their role have been often contradictory (Davis and Gregoriadis, 1987; Bakouche and Gerlier, 1986; Shek, 1984; van Rooijen and van Nieuwmegen, 1980) or unconfirmed (Francis et al, 1985; Shek, 1984; Latif and Bachhawat, 1987).

Whereas it is likely that some of these contradictions are the result of differences in experimental protocols, others could be explained as the result of the anticipated diversity of interactions between membrane- or water-soluble antigens (used in conjunction with liposomes) and APC in situ. Interactions may include supply of free antigen to APC following disinteg-ration of liposomes locally, direct antigen transfer by intact liposomes to the APC or, after (liposomal) antigen catabolism intracellularly, migration of fragmented forms to the APC membrane. This diversity of events is likely to be further augmented by variations in liposomal structural character-istics. For instance, early studies (Bakouche and Gerlier, 1986; Kinsky, 1978) on the effect of bilayer fluidity on the immune responses to membrane-soluble antigens indicated that liposomes made of phospholipids (eg. DSPC) with a Tc above 37°C elicit strong antibody responses to the antigens. The reverse, however, was observed for water soluble antigens (eg. tetanus toxoid): strong responses were now seen with liposomes made of phospholipids with low Tcs, whilst those obtained with DSPC liposomes were nil or minimal. The differential effect of DSPC on responses to the two types of antigen was tentatively attributed (Davis and Gregoriadis, 1987) to the direct transfer of membrane antigens from the liposome carrier to the plasma membranes of APC where they may associate with MHC molecules, without being first pro-cessed. In this respect, it is of interest that the immunogenicity of a small poliovirus peptide, expected to interact directly with the MHC on the APC membranes was also greatest when administered in DSPC liposomes (Xiao et al, 1989). Soluble antigens such as tetanus toxoid are, in contrast, probably transported directly into the APC's interior where they are pro-cessed prior to the presentation of their fragments on the cell membranes. It is conceivable that both processing and presentation of soluble antigens could be inhibited or interfered with by DSPC or other high melting phospho-lipids (Davis and Gregoriadis, 1987).

Control of liposomal adjuvanticity by bilayer fluidity becomes even more complex by the participation of at least one other factor: both primary (Davis and Gregoriadis, 1989) and secondary (Davis and Gregoriadis, 1987) responses induced by PC liposomes against their antigen (tetanus toxoid) content are improved considerably when preparations with a high (eg. $2x10^3$) liposomal lipid to antigen mass ratio are used. There is, however, an equally improved response to toxoid for DSPC liposomes with similarly high ratios (Davis and Gregoriadis, 1987). As discussed elsewhere (Davis and Gregoriadis, 1987), it is possible that the relatively large amount of lipid (DSPC) involved contributes to a slower and more efficient supply of antigen to APC thus overriding any inhibitory effect of DSPC on the immunogenicity of antigen taken up by the cells in the liposome form. For liposomes with even higher phospholipid to antigen mass ratios (eg. $>10^4$), there was no immunoadjuvant activity, presumably because the concentration of antigen in individual vesicles was too low for it to be immunogenic (Davis and Gregoriadis (1987). It is thus essential that, when comparisons of antibody responses produced by different liposomal-antigen formulations are made, liposomal lipid to antigen mass ratios must be similar. Failure to observe this requirement may explain (Davis and Gregoriadis, 1987) conflicting results (Shek, 1984; van Rooijen and van Nieuwmegen, 1980) from studies on the effect of antigen localization within liposomes on adjuvanticity.

It is thus reasonable to suggest that a number of conditions, including that of an appropriate lipid to antigen mass ratio, are instrumental in influencing the balance among the various mechanisms (as already discussed) of liposomal antigen supply to APC in vivo. Such balance is likely to reflect primarily the structural profile of a given liposome vaccine formulation. For example, the rate and extent of liposome degradation in situ and subsequent antigen release depend on vesicle lipid composition and average size. Composition and size will, in turn, also control vesicle uptake by the lymphatics, mode of interaction with APC as well as intracellular fate. Moreover, such events may also be influenced in a variety of ways by different antigens because of variability in their spatial arrangement within liposomes and the way they may interact with their lipid components. Consequences of such complex scenarios of adjuvanticity in vivo in terms of its study and modification in vitro have been discussed elsewhere (Francis and Clarke, 1989).

LIPOSOMAL ADJUVANTICITY FOR SMALL PEPTIDES

Work on liposomal adjuvanticity for small peptides has been discussed briefly elsewhere (Gregoriadis, 1990) and includes studies on three synthetic peptides which were found to generate immunological memory (Xiao et al, 1989; Alving et al, 1986; Richards et al, 1988; Francis et al, 1985). Of these peptides, hepatitis B virus pre-S (Steward et al, 1988) and foot-and-mouth disease virus VP1 (Francis and Clarke, 1989) peptides contain B- and T-cell epitopes and the same may be true for the poliovirus 3VP2 peptide against which immunological memory was produced (Xiao et al, 1989) on injection in the liposome form. It seems that by entrapping small peptides into liposomes of an appropriate lipid composition and size, the need for a carrier protein and associated problems (Francis and Clarke, 1989) can be dispensed with. It is also likely that the presence of both B- and T-cell epitopes on a peptide is necessary for liposomes to induce memory for the B-epitope. In the absence of a T-epitope, liposomal B-peptides are expected to behave as haptens (T-cell-independent antigens) producing IgM response. On the other hand, a liposome-entrapped B-cell epitope may receive help from a co-entrapped T-cell epitope to produce an IgG response. T and B epitopes could then interact with MHC class II molecules on APC and B cells respectively in the trimolecular fashion proposed (Francis and Clarke, 1989) for peptide conjugates incorporating both epitopes, to produce immunological memory. Indeed, recent work from our laboratory (in collaboration with Dr. Z. Wang and Dr. M. Francis, Pitman-Moore, Harefield, Middlesex, UK) suggests that this is, indeed, the case. For instance, a hepatitis B surface antigen liposome-entrapped pre-S$_2$ peptide, which by itself is unable to induce an IgG response in the strain of mice used, does so when co-entrapped in the same liposomes with an S peptide.

CONCLUSIONS

Results from a large number of animal immunization studies indicate that liposomes produce humoural and cell-mediated immunity to a wide range of antigens. Liposomal adjuvanticity appears to depend on several of the system's structural characteristics which are known to determine its fate in vivo and, thus, the mode of antigen interaction with APC. Further amplification of adjuvanticity can be obtained by receptor mediated targeting to macrophages or the presence of other adjuvants and cytokines. In addition to their function as immunoadjuvants, liposomes can also serve as a carrier for independently co-entrapped B- and T-cell epitopes, thus eliminating the need for a carrier protein.

Acknowledgements

The author's work cited in this chapter was supported by Medical Research Council project grants, Wellcome Biotechnology, Beckenham and the British Council.

REFERENCES

Allison, A.C. and Byars, N.E., 1986, An adjuvant formulation that selectively elicits the formation of antibodies of protective isotypes and of cell-mediated immunity, J.Immunol.Meth., 95:157.

Allison, A.C. and Gregoriadis, G., 1974, Liposomes as immunological adjuvants, Nature, 252:252.

Almeida, J.D., Brand, C.M., Edwards, D.C. and Heath, T.D., 1975, Formation of virosomes from influenza virus subunits and liposomes, Lancet, 2:889.

Alving, C.R., Richards, R.L., Moss, J., Alving, L.I., Clements, R.D., Shiba, T., Kotani, S., Wirtz, R.A. and Hockmeyer, W.T., 1986, Effectiveness of liposomes as potential carriers of vaccines. Application to cholera toxin and human malaria sporozoite antigen, Vaccine, 4:166.

Bakouche, Q. and Gerlier, D., 1986, Enhancement of immunogenicity of tumour virus antigen by liposomes. The effect of lipid composition, Immunology, 57:219.

Blume, G. and Cevc, G., 1990, Liposomes for the sustained drug release in vivo, Biochim.Biophys.Acta, 1029:91.

Bomford, R., 1985, Adjuvants, in: "Animal Cell Biotechnology", Spier, R. and Griffiths, B. (eds.), Vol. 2, Academic Press, London.

Dailey, M.O. and Hunter, R.L., 1977, Induction of cell-mediated immunity to chemically-modified antigens in guinea pigs. 1. Lipid-conjugated protein antigens, J.Immunol. 118:957.

Davis, D. and Gregoriadis, G., 1987, Liposomes as adjuvants with immuno-purified tetanus toxoid: Influence of liposomal characteristics, Immunology, 61:229.

Davis, D. and Gregoriadis, G., 1989, Primary immune response to liposomal tetanus toxoid in mice: the effect of mediators, Immunology, 68:277.

Davis, D., Davies, A. and Gregoriadis, G., 1987, Liposomes as adjuvants with immunopurified tetanus toxoid: The immune response, Immunol.Lett., 14:341.

Francis, M.J. and Clarke, B.E., 1989, Peptide vaccines based on enhanced immunogenicity of peptide epitopes presented with T-cell determinants or hepatitis B core protein, Meth.Enzymol. 178:659.

Francis, M.J., Fry, C.M., Rowlands, D.J., Brown, F., Bittle, J.L., Houghten, R.A. and Lerner, R.A., 1985, Immunological priming with synthetic peptides of foot-and-mouth disease virus, J.Gen.Virol., 66:2347.

Garcon, N., Gregoriadis, G., Taylor, M. and Summerfield, J., 1988, Targeted immunoadjuvant action of tetanus toxoid-containing liposomes coated with mannosylated albumin, Immunology 64:743.

Gregoriadis, G., 1976, The carrier potential of liposomes in Biology and
    Medicine, New Engl.J.Med., 295:704 and 765.
Gregoriadis, G. (ed.), 1984, "Liposome Technology", Vols. 1-3, CRC Press
    Inc., Boca Raton.
Gregoriadis, G. (ed.), 1988a, "Liposomes as Drug Carriers: Recent Trends and
    Progress", John Wiley and Sons, Chichester.
Gregoriadis, G., 1988b, Fate of injected liposomes: Observations on
    entrapped solute retention, vesicle clearance and tissue distribution
    in vivo, in: "Liposomes as Drug Carriers: Recent Trends and
    Progress", Gregoriadis, G., (ed.), John Wiley and Sons, Chichester.
Gregoriadis, G., 1990, Immunological adjuvants: A role for liposomes,
    Immunology Today, 11:89.
Gregoriadis, G., Davis, D. and Davies, A., 1987, Liposomes as immunological
    adjuvants: Antigen incorporation studies, Vaccine, 5:143.
Gregoriadis, G., Allison, A.C. and Poste, G. (eds.), 1989, Immunological
    Adjuvants and Vaccines, Plenum, New York.
Gregoriadis, G., Tan, L. and Xiao, Q., 1989, The immunoadjuvant action of
    liposomes: Recent Progress, in: "Immunological Adjuvants and
    Vaccines", Gregoriadis, G., Allison, A.C. and Poste, G. (eds.),
    Plenum, New York.
Gregoriadis, G., da Silva, H. and Florence, A.T., 1990, A procedure for the
    efficient entrapment of drugs in dehydration-rehydration liposomes
    (DRV), Int.J.Pharmaceutics, 65:235.
Kahl, K.L., Scott, C.A., Lelchuk, R., Gregoriadis, G. and Liew, F.Y., 1989,
    Vaccination against murine cutaneous leishmaniasis using L. Major
    antigen/liposomes: Optimization and assessment of the requirement
    for intravenous immunization, J.Immunol., 142:4441.
Kinsky, S.C., 1978, Immunogenicity of liposomal model membranes, Ann.N.Y.
    Acad.Sci., 308:111.
Kirby, C. and Gregoriadis, G., 1984, Dehydration-rehydration vesicles (DRV):
    A new method for high yield drug entrapment in liposomes,
    Biotechnology, 2:979.
Kraaijeveld, C.A., Schilham, M., Jansen, J., Benaissa-Trouw, B., Harmsen,
    M., van Houte, A.J. and Snippe, H. (1984), The effect of liposomal
    charge on the neutralizing antibody response against inactivated
    encephalomyocarditis and Semliki Forest viruses, Clin.exp.Immunol.,
    56:509.
Latif, N.A. and Bachhawat, B.K., 1987, The effect of surface-coupled antigen
    of liposomes on immunopotentiation, Immunol.Lett., 15:45.
Lopez-Berestein, G. and Fidler, I.J. (eds.), 1989, "Liposomes in the Therapy
    of Infectious Diseases and Cancer", Alan R. Liss Inc., New York.
Manesis, E.K., Cameron, C.H. and Gregoriadis, G., 1979, Hepatitis B surface
    antigen-containing liposomes enhance humoral and cell-mediated
    immunity to the antigen, FEBS Lett., 102:107.
Mayhew, E., Lazo, R., Vail, W.J., King. J and Gree, A.M., 1984, Character-
    ization of liposomes using a microemulsifier, Biochim.Biophys.Acta,
    775:169.
Papahadjopoulos, D., Allen, T., Gabizon, A., Mayhew, E., Matthay, K., Huang,
    S.K., Lee, K.D., Woodle, M.C., Lasic, D.D., Redemann, C. and
    Martin, F.J., 1992, Sterically stabilized liposomes: pronounced
    improvements in blood clearance, tissue disposition and therapeutic
    index of encapsulated drugs against implanted tumours, Proc.Nat.
    Acad.Sci.USA (in press).
Richards, R.L., Hayre, M.D., Hockmeyer, W.T. and Alving, C.R., 1988,
    Liposomal lipid A and aluminium hydroxide enhance the immune
    response to a synthetic malaria sporozoite antigen, Infect.Immun.,
    56:682.
van Rooijen, N. and van Nieuwmegen, R., 1980, Liposomes in immunology:
    Evidence that their adjuvant effect results from surface exposition
    of the antigen, Cell Immunol., 49:402.

Senior, J., Crawley, J.C.W. and Gregoriadis, G., 1985, Tissue distribution of liposomes exhibiting long half-lives in the circulation after intravenous injection, Biochim.Biophys.Acta, 839:1.

Senior, J., Delgado, C., Fisher, D., Tilcock, C. and Gregoriadis, G., 1991, Influence of surface hydrophilicity of liposomes on their interaction with plasma proteins and clearance from the circulation: Studies with polyethylene glycol-coated vesicles, Biochim.Biophys. Acta, 1062:77.

Shek, P.N., 1984, Application of liposomes in immunopotentiation, in: "Immunotoxicology", Mullen, P.W. (ed.), Springer-Verlag, Berlin.

Steward, M., Sisley, B., Stanley, C., Brown, S. and Howard, C., 1988, Immunity to hepatitis B: analysis of antibody and cellular responses in recipients of plasma-derived vaccine using semisynthetic peptides mimicking S and pre-S regions, Clin.Exp.Immunol., 71:19.

Su, D. and van Nieuwmegen, R., 1989, The role of macrophages in the immunoadjuvant action of liposomes: effects of elimination of splenic macrophages on the immune response against intravenously injected liposome-associated albumin antigen, Immunology, 66:466.

Tan, L. and Gregoriadis, G., 1989a, The effect of positive surface charge of liposomes on their clearance from blood and its relation to vesicle lipid composition, Biochem.Soc.Trans., 17:690.

Tan, L. and Gregoriadis, G., 1989b, The effect of interleukin-2 on the immunoadjuvant action of liposomes, Biochem.Soc.Trans., 17:693.

Tan, L., Loyter, A. and Gregoriadis, G., 1989, Incorporation of reconstituted influenza virus envelopes into liposomes: Studies of immune response in mice, Biochem.Soc.Trans. 17:129.

Tumer, A., Kirby, C., Senior, J. and Gregoriadis, G., 1983, Fate of cholesterol-rich unilamellar liposomes containing ^{111}In-labelled bleomycin after subcutaneous injection into rats, Biochim.Biophys. Acta, 760:119.

Walden, P., 1988, Antigen presentation by liposomes as model system for T-B cell interaction, Eur.J.Immunol., 18:1851.

Xiao, Q., Gregoriadis, G. and Ferguson, M., 1989, Immunoadjuvant action of liposomes for entrapped poliovirus peptides, Biochem.Soc.Trans., 17:695.

TARGETING PROTEINS TO ANTIGEN-PRESENTING CELLS AND INDUCTION OF CYTOKINES
AS A BASIS FOR ADJUVANT ACTIVITY

Anthony C. Allison and Noelene E. Byars

Institute of Immunology and Biological Sciences, Syntex
Research, Palo Alto, CA 94304, U.S.A.

INTRODUCTION

The need for improved and new vaccines is evident. While live virus
vaccines have been very useful, they can produce severed infections in
persons whose immune systems are compromised because of congenital
deficiency, malnutrition or infections (notably HIV). If subunit
antigens can elicit equivalent protection they are preferred.
Preparation of antigens by recombinant technology has opened up the
possibility of a new generation of vaccines, but a safe and efficacious
adjuvant is a necessary component. The only adjuvant approved for human
use is aluminum hydroxide or phosphate. While alum is an effective
adjuvant for bacterial toxoids it has limited efficacy in other
situations, for reasons that will be discussed below.

With subunit vaccines in experimental animals Freund's complete
adjuvant (FCA) is often required to elicit protection (Berman et al.,
1985; Morgan et al., 1989). The aim of our research program was to
develop an adjuvant formulation with the efficacy of FCA but without its
unacceptable side effects. The most serious side effect of FCA is
granulomas at injection sites, but tuberculin hypersensitivity is also
undesirable. Our strategy was to replace the components of FCA with
acceptable alternatives and to define efficacy more rigorously than had
been done in the past.

MURAMYL DIPEPTIDE ANALOGS

FCA consists of killed mycobacteria in a water-in-oil emulsion
comprising mineral oil and Arlacel A; Freund's incomplete adjuvant (FIA)
has the same composition but lacks the mycobacteria. Most antigens
administered in FIA elicit antibody responses but not delayed-type
hypersensitivity (DTH), whereas when the same antigens are given in FCA
DTH is also elicited. Guinea pigs immunized with antigens in FIA produce
antibodies mainly of the γ_2 isotype whereas FCA increases the formation
of antibodies on the γ_1 isotype (White, 1976). One objective of adjuvant
research has been to define an acceptable alternative to the mycobacteria
in FCA. An importan' step was taken by Ellouz et al. (1974), who showed
that the minimal adjuvant-active component of mycobacteria was N-
acetylmuramyl-L-alanyl-D-isoglutamine (alanyl-MDP). When antigens were

Targeting of Drugs 3: The Challenge of Peptides and Proteins
Edited by G. Gregoriadis et al., Plenum Press, New York, 1992

69

administered to guinea pigs in FIA + alanyl-MDP, the responses were like those elicited using FCA.

Substitution of mycobacteria by synthetic MDP in adjuvants avoided tuberculin hypersensitivity but not other problems. Biologically active bacterial products, including MDP and lipopolysaccharide (LPS), have common properties manifested to varying degrees in different species of animals. MDP and LPS and pyrogenic, which precludes use in vaccines. They can stimulate nonspecific immunity against infections and tumours, and they can induce arthritis and anterior uveitis. The most familiar example is Reiter's syndrome associated with some Gram-negative bacterial infections in genetically predisposed humans, often of the HLA-B27 haplotype (Geczy et al., 1983): the syndrome includes arthritis and anterior uveitis. An experimental model of arthritis is adjuvant arthritis in the rat: normally this is elicited by FCA, but alanyl-MDP in FIA can have the same effect (Nagao and Tanaka, 1980). An experimental model of uveitis is the rabbit: intravenous injection of LPS or alanyl-MDP induces an early increase in vascular permeability in the eye, shown by passage of fluoresceinated macromolecules into the anterior chamber; leukocytes adhere to endothelium and migrate into the anterior uvea (Waters et al., 1986).

Our first objective was to identify a synthetic analog of MDP with potent adjuvant activity but with low pyrogenicity and capacity to induce arthritis and uveitis. More than 120 MDP analogs were tested for adjuvant activity in guinea pigs, using FIA as vehicle and arsanilic tyrosine and either bovine serum albumin or ovalbumin as model antigens. Both humoral and cell-mediated responses were measured. The parent alanyl-MDP in FIA was the positive control, and FIA was the negative control. We selected only those MDP analogues which induced higher antibody titres to albumin, and stronger DTH skin reactions to both arsanilic tyrosine and albumin, than did the parent compound. Dose-response assays were then performed with the more promising analogues. Twenty of these were tested in a series of models, including carbon clearance, measurements of anti-infective activity in mice against pseudomonas and candida, induction of adjuvant arthritis in rats, pyrogenicity in rats and rabbits, induction of uveitis in rabbits, and induction IL-1 synthesis by guinea pig macrophages and human monocytes in culture.

From the data obtained in all these assays threonyl-MDP was selected as the most suitable analogue (Allison and Byars, 1986, Fig. 1). It has excellent adjuvant activity, is pyrogenic and uveogenic only at very high

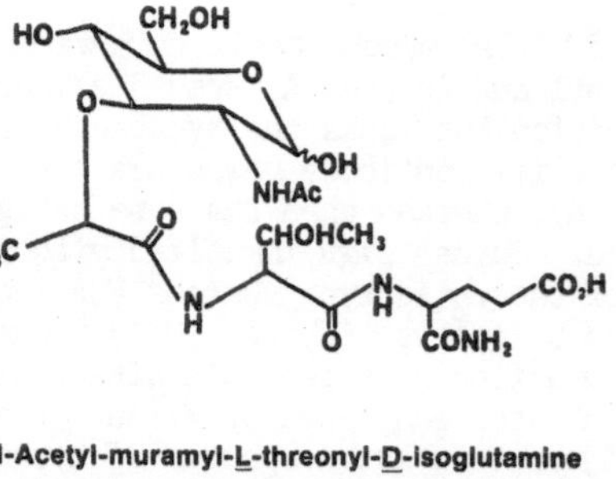

N-Acetyl-muramyl-L-threonyl-D-isoglutamine

Fig. 1. Structure of N-acetylmuramyl-L-threonyl-D-isoglutamine.

doses (Waters et al., 1986), and does not induce adjuvant arthritis. The lack of toxicity shown by the threonyl analogue is in contrast to the parent alanyl-MDP, as well as many synthetic analogues, for example desmethylalanyl-MDP and aminobutyryl-MDP (Waters et al., 1986). Some of the muramyl peptide analogues appear to be broad-based, non-specific immune stimulants, like the parent alanyl-MDP. A notable example is muramyltripeptide phosphatidylethanolamine. However, the activity of threonyl-MDP is largely confined to adjuvant activity, since it does not stimulate carbon clearance in mice and shows no anti-infective activity (Fraser-Smith et al., 1982). The adjuvant-specific activity of threonyl-MDP can be regarded as advantageous since it reduces the likelihood of unwanted side effects.

The reason for the difference in biological activities of alanyl-MDP and threonyl-MDP is not fully understood. However, nuclear magnetic resonance analysis suggests that intramolecular hydrogen bonding restricts the conformation of alanyl-MDP (Fermandjian et al., 1987). The additional hydroxyl group in threonyl-MDP can change the hydrogen bonding pattern and configuration of the molecule.

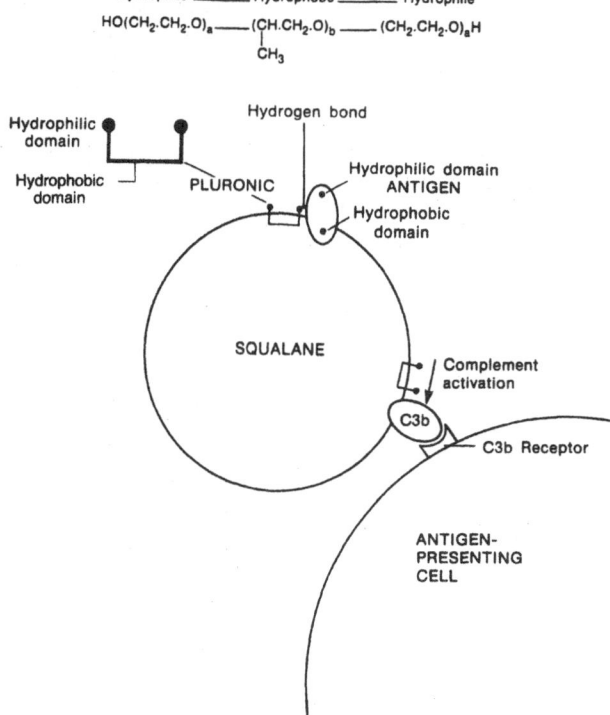

Fig. 2. Above structure of Pluronic[R] copolymer L121. Blocks of polyoxyethylene (POE) flank a central block of polyoxypropylene (POP). Below is a diagram of a squalane microsphere showing L121 at the surface interacting with a model antigen, and with the activated third component of complement (C3b).

SQUALANE POLYOXYETHYLENE-POLYPROPYLENE COPOLYMER EMULSION

MDP analogs are effective adjuvants when injected with antigens in FIA but not in saline. These and other observations suggest that partitioning of antigens at a lipid-water interphase improves immunogenicity. Our next task was to develop an acceptable substitute for FIA. Mineral oil itself was found to produce granulomas in humans, and there has been concern about carcinogenicity (Potter and Boyce, 1962; Murray et al., 1972). We also investigated the choice of emulsion required for antigen presentation. When water-in-oil emulsions such as FCA and FIA are injected, depots of the bulk oil phase remain at the injection site and become infiltrated with macrophages, leading to granulomas. We preferred oil-in-water emulsions, since the oil spherules can migrate from the injection site into lymphatics and carry antigens to lymph nodes of the drainage chain. The bulk liquid phase of the injected vehicle rapidly dissipates. Depots of antigen are formed on antigen-presenting cells in lymphoid tissue, not at the injection site.

Squalene and squalane were found under the right circumstances to produce oil-in-water emulsions. Squalene is a precursor of cholesterol and an abundant body constituent. However, it is unsaturated and susceptible to autoxidation, which is inconvenient in pharmaceutical formulations. Squalene can be used with a reducing agent, but it is simpler to use the saturated form, squalane, which is also a body constituent (Gosselin et al., 1976) and is generally regarded by regulatory authorities as safe in pharmaceutical and cosmetic formulations. Squalane was therefore selected as a replacement for the mineral oil in FCA. The emulsifying surface-active agent in FCA, Arlacel A, has also come under suspicion as a potential carcinogen (Murray et al., 1972), so that another surface-active agent was required. A lead came from observations of Hunter et al. (1981) that synthetic polyoxyethylene-polyoxypropylene (Pluronic[R]) copolymers, when used with mineral oil, have adjuvant activity. We found that squalene or squalane, L121 copolymer and a small amount of Tween 80 in phosphate-buffered saline (PBS) can be emulsified by microfluidization to produce a remarkably stable emulsion. The combination of threonyl-MDP with the squalane-L121 emulsion is termed Syntex Adjuvant Formulation (SAF). The pre-formed emulsion is gently mixed with a solution of antigen and threonyl-MDP to produce the final vaccine.

The antigen retains its native structure, since it is not subjected to the mixing and shearing forces of the emulsification process. This gentle mixing of pre-formed emulsions and antigen-MDP solution is a totally different preparation method from that used for making Freund's-type antigen emulsions. When antigens are injected in FCA, high titers of antibodies recognizing internal determinants (exposed by denaturation) are observed, which is not the case when our adjuvant formulation is used (Kenney et al., 1989). Furthermore, the vaccines made with SAF have the consistency of milk, and are therefore easy to inject. The emulsion is very stable, and can be stored at room temperature or in the refrigerator (4°C) for months; it is even stable when frozen.

We postulate that the oil microspheres in the SAF emulsion consist of squalane, with the L121 on the surface. The L121 copolymer consists of a central block of polyoxypropylene, which is hydrophobic, flanked on either side by smaller blocks of the hydrophilic polyoxyethylene (Fig. 2). We suggest that the polymer allows the oil microspheres to retain antigens on their surfaces (Fig. 3), partly because of the amphipathic character of the antigens and partly by hydrogen bonding to polyoxyethylene. Complement components present in tissue fluids are also bound. As a result, complement is activated via the alternative pathway.

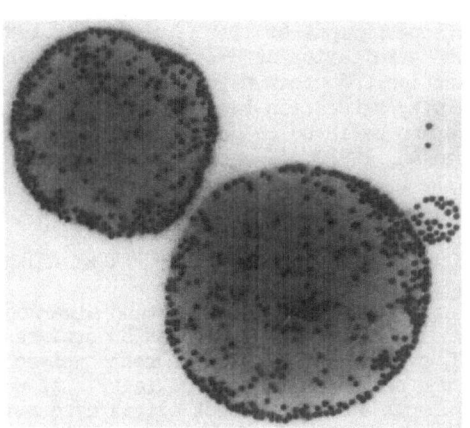

Fig. 3. Electron micrograph showing ovalbumin labelled with
colloidal gold on the surface of squalane emulsion
particles (x30,000).

The activated third component of complement, C3b, on the surface of the particles facilitates their targeting to follicular dendritic cells, which have receptors for C3b. The oil microspheres thus carry antigen to the surface of antigen-presenting cells, where the antigen molecules can be endocytosed, then subsequently re-expressed on the cell surface in association with MHC molecules for presentation to lymphocytes with antigen-specific receptors.

ANTIGEN-PRESENTING CELLS

Bone marrow-derived dendritic cells (DC) migrate from bone marrow to the spleen and to various tissues of the body. The tissue-specific stage of their life history is exemplified by Langerhans cells in the skin. The next stage is migration through afferent lymphatics (where they are termed veiled cells) to the T-dependent areas of lymph nodes of the drainage chain. That stage is known as interdigitating cells. DC are potent antigen-presenting cells (APC and are especially effective in recruiting T-cells into immune responses (Metlay et al., 1990).

The follicular dendritic cell (FDC) is another major APC. As their name implies, they are found in follicles in the areas of lymphoid tissues rich in B-lymphocytes. Complement-fixing antigen-antibody complexes are bound by FDC and antigen is retained in association with these cells for long periods of time (Tew et al., 1989). This is important for the elicitation of B-memory, which is required for effective vaccination. Another requirement is for high-affinity antibodies: affinity maturation involves somatic hypermutation of B-cells, a process now believed to be confined to the germinal centres of follicles (MacLennan, 1991). B-cells can function as APC for T-cells, a process especially important in secondary responses.

SYNTEX ADJUVANT FORMULATION (SAF) IN LABORATORY ANIMALS

Injection of a variety of antigens into mice and guinea pigs has elicited cell-mediated immunity (CMI) and humoral responses. CMI is assessed by DTH, proliferative cellular responses to antigens, production of lymphokines and cell-mediated cytotoxicity. It was traditionally believed that subunit antigens cannot elicit cytotoxic T-cells. However, we have found that vaccination of guinea pigs with a recombinant surface antigen of herpes simplex virus (gDt) in SAF elicits T-cells able to use target cells expressing the antigen in a MHC class II-restricted fashion (Table 1). The envelope protein of HIV injected into mice with ISCOMS elicits class I-restricted $CD8^+$ T-cells cytotoxic (Takahashi et al., 1990) so live viruses are unnecessary to elicit cytotoxic responses. Cytotoxic T-cells do not necessarily protect by killing virus-infected cells. When they meet cells expressing virus-specific antigens, cytotoxic T-cells release lymphokines such as IFN-γ, which has anti-viral activity (Morris et al., 1982).

Lymphokines produced by T-lymphocytes influence the isotype of antibodies produced. IL-4 increases the formation of murine IgG1 antibodies and appears to be an absolute requirement for the formation of IgE antibodies; in contrast, IFN-γ augments the production of IgG2a antibodies in the mouse (Finkelman et al., 1990). The isotypes of antibodies elicited are of interest for two reasons: first, because of

Table 1. Induction, by vaccination with recombinant subunit (gD) antigen of HSV-2 in SAF, of CD4[+], MHC-class II restricted cytotoxicity (data of A.R. Hayward and A.C. Allison).

Effector	No. of Animals	Antigen[a]	Antibody	% Lysis[b]	
Strain 2	6	0	0	5.6	4
	6	gD	0	36	8
	4	gD	anti-Ia	12	6
	4	gD	anti-CD4	10	3
Strain 13	4	0	0	6	3
	4	gD	0	7	5
F1 (2x13)	3	0	0	6	5
	3	gD	0	28	5

[a] 1 µg gD in SAF administered intramuscularly. Controls received SAF alone.

[b] Methods described in Hayward et al. (1991).

the role of subsets of T-cells and their mediators in augmenting the formation of antibodies of different isotypes, and secondly because of the protective role of antibodies of different isotypes. For protection in some situations, isotypes have little influence, e.g. antibodies of any isotype, provided that they are of sufficiently high affinity, can bind bacterial toxins and prevent them from combining with receptors on target cells. However, antibodies of certain isotypes activate complement well and bind to $Fc_{\gamma}I$ receptors on leukocytes (Unkeless et al., 1988). Binding to $Fc_{\gamma}I$ receptors opsonizes bacteria for phagocytosis and mediates antibody-dependent cytotoxicity (ADCC). In the mouse, IgG2a antibodies fulfil this role and in humans IgG1 antibodies do so most efficiently. Studies with isotype-switch variants of murine monoclonal antibodies show that IgG2a antibodies more efficiently mediate ADCC in vitro, and are more protective against tumours in vivo, than are antibodies of other isotypes (Steplewski et al., 1985; Kaminski et al., 1986). Studies with "reshaped" human antibodies who the IgG1 isotype to be superior to other isotypes in mediating ADCC (Reichmann et al., 1988). Augmenting the production of IgG2a antibodies in the mouse, and IgG1 antibodies in humans, can therefore broaden the range of protective responses to a vaccine antigen.

Antigens in SAF increase the production of IgG2a antibodies in mice (Kenney et al., 1989; Byars et al., 1990, 1991). This is attributed to the activation of cytotoxic T-cells, which can produce IFN-γ, as well as the preferential activation of TH1 cells, which also produce IFN-γ (Mosmann and Coffman, 1989). In contrast, other adjuvants, such as alum, preferentially stimulate the production of IL-4 and of IgG1 antibodies in the mouse (Bomford, 1989). Observations supporting the interpretation that SAF increases the number of cells producing IFN- in lymph nodes draining sites of antigen injection, as well as in the spleen, will be presented elsewhere.

From the practical point of view, it is fortunate that elicited cytotoxic T-cell and TH1 helper cell responses both augment the protective efficacy of a vaccine: favourable cell-mediated and humoral immune responses are not mutually exclusive. We have also found that high-affinity antibodies against several antigens, including some of low immunogenicity, can be raised by subcutaneous injection of SAF; to raise monoclonal antibodies, intraperitoneal injection in FCA, followed by FIA, is unnecessary (Kenney et al., 1989).

USE OF SAF IN VACCINES

Our first application of SAF was with inactivated feline leukaemia virus, resulting in the development of the first effective vaccine against this virus (Braemer et al., 1984). SAF was also successfully used with inactivated viruses to immunize rhesus monkeys against simian AIDS virus (Marx et al., 1986) and simian immunodeficiency virus (Murphey Corb et al., 1989). More recently SAF has been used with recombinant envelope antigens of HIV to immunize chimpanzees (Girard et al., 1991).

Other studies with subunit antigens were designed to address particular questions. One is whether SAF can overcome effects of age and genetic restriction on immune responses. Young adult humans and experimental animals respond optimally to vaccines, while responses of newborns and old individuals are lower and less consistent. For example, influenza vaccine is recommended for persons aged 65 years or over, but less than 30% of old human recipients of the vaccine show seroconversion (Arden et al., 1986). Influenza virus hemagglutinin (HA) is administered in saline because alum precipitation does not improve immune responses

(Nicholson et al., 1979). We found that in young and old mice responses to HA are low and variable; administering the antigen in SAF improves responses in all age groups and makes them more consistent (Byars et al., 1990). Vaccinating newborn children against hepatitis B virus can prevent the perinatal transmission of the virus, which increases the risk of long-term carriage, cirrhosis and hepatocellular carcinoma. Again, age limits the response. Even in young adults three doses of serum-derived or recombinant surface antigen (alum-adjuvanted) are required to elicit seroconversion in the great majority of recipients; about 5% are low responders, apparently for genetic reasons. Fewer old humans respond to the standard three-dose vaccination schedule. When mice and guinea pigs were used as experimental models SAF was clearly superior to alum (Byars et al., 1991). With SAF the dose of recombinant HBsAg required to elicit responses regarded as protective could be decreased at least tenfold (important for an expensive antigen), two doses were sufficient, and antibodies could be elicited even in genetically low-responder mice. If these findings can be extended to humans, currently used vaccines will be improved.

The next question addressed is whether subunit antigens of herpes viruses can be used to elicit protection. This question is of academic interest because cell-mediated immune responses are believed to play a major role in limiting these infections, once they are established. Epstein-Barr virus (EBV) transforms B-lymphocytes and stimulates polyclonal proliferation; T-lymphocytes can restrict the outgrowth of the transformed B-lymphocytes (Rickinson et al, 1984). In susceptible subhuman primates, cottontop tamarins (Sanguinus oedipus) EBV produces a fatal lymphoproliferative syndrome. Vaccination with a major envelope glycoprotein of EBV (gp360) in SIV protected tamarins against virus challenge (Morgan et al., 1989). In this model the same antigen in alum was ineffective, and acceptable vaccinia constructs showed limited efficacy.

A model of genital herpes is produced by vaginal infection of guinea pigs with HSV-2 (Table 2). The virulent strain of virus used produces severe vaginal lesions; in most animals it infects dorsal root nerve ganglia, the source of recurrent infections, and it produces lethal encephalitis in some recipients. Following vaccination with a recombinant surface antigen (gDt) in SAF, intravaginal infection elicited small lesions, virus could not be recovered from dorsal root ganglia (with one exception) and there were no lethal infections. These studies show the feasibility of herpes virus vaccines in experimental animals. In this system alum is ineffective. SAF has also been used to elicit anti-idiotypic responses in mice. These provided protection against B-lymphomas expressing the idiotype (Campbell et al., 1989).

PROSPECTS

The adjuvant components were not found to be mutagenic, and extensive toxicology studies suggested that they produce minimal reactions at injection sites and are otherwise acceptable. Current trials show SAF to be useful for eliciting anti-idiotypic responses in humans, without limiting toxicity. Obtaining regulatory approval for the use of the adjuvant should be a high priority, since it could be a component of much-needed vaccines (e.g. against HIV). Either SAF or a similar formulation should replace FCA for routine laboratory animal immunization. Our Squalene or squalane oil-in-water emulsions have been adopted by others using other muramyl peptide analogs or monophosphoryl lipid A (Ribi et al, 1987), which shows the general utility of the strategy.

Table 2. Effect of vaccination with recombinant gD of HSV-2 in SAF on the course of primary vaginal lesions and establishment of latency in dorsal root ganglia (observations of N. Byars, B. Fraser-Smith and A. C. Allison).

Prior Vaccination	No. of Animals	% with Vaginal Lesions	Mean Lesion Score	% Ganglia Infected
SAF (no antigen)[a]	22	100	2.66	64
SAF + gD[a]	20	45[b]	0.13[c]	5[b]

[a] Intramuscular administration of SAF (adjuvant) alone or with 1 μg recombinant gD

[b] p<0.05 (Fisher Exact Probability)

[c] p<0.05 (ANOVA)

REFERENCES

Allison, A. C., and Byars, N. E., 1986, An adjuvant formulation that selectively elicits the formation of antibodies of protective isotypes and cell-mediated immunity, J. Immun. Methods, 95:157.

Arden, N. H., Patriarca, P. A., and Kendal, A. P., 1986, Experiences in the use and efficacy of influenza vaccine in nursing homes, in: "Options for Control of Influenza", A. P. Kendal, and P. A. Patriarca, eds., Alan R. Liss, New York.

Berman, P. W., Gregory, T., Crase, P., and Laswky, L. A., 1985, Protection from genital herpes simplex type 2 infection with cloned glycoprotein D., Science, 227:1490.

Bomford, R., 1989, Aluminium salts: perspectives in their use as adjuvants, in: "Immunological Adjuvants and Vaccines", G. Gregoriadis, A. C. Allison and G. Poste, eds., Plenum Press, London.

Braemer, A., Peterson, M., Renneke, G., Bass, E., Allison, A. C., Byars, N. E., and Fraser, D., 1984, Effect of inactivated FeLV vaccines on the development of persistent viremia, Proc. 65th Conf. Res. Workers in Animal diseases (E.M. Kohler) (p. 10).

Byars, N. E., and Allison, A. C., 1987, Adjuvant formulation for use in vaccines to elicit both cell-mediated and humoral immunity, Vaccine, 5:223.

Byars, N. E., Allison, A. C., Harmon, M. W., and Kendal, A. P., 1990, enhancement of antibody responses to influenza B virus hemagglutinin by use of a new adjuvant formulation, Vaccine, 8:49.

Byars, N. E., Nakano, G., Welch, M., Lehman, D., and Allison, A. C., 1991, Improvement of Hepatitis B vaccine by the use of a new adjuvant, Vaccine, 9:309.

Campbell, M. J., Esserman, L., Byars, N. E., Allison, A. C., and Levy, R., 1989, Development of a new therapeutic approach to B-cell malignancy. The induction of immunity by the host against cell surface receptor on the tumour, Int. Rev. Imm., 4:251.

Ellouz, F., Adam, A., Ciorbaru, R., and Lederer, E., 1974, Minimal structural requirements for adjuvant activity of bacterial peptidoglycans, Biochem. Biophys, Res. Commun., 59:1317.

Fermandijian, S., Perly, B., Level, M., and Lefrancier, P., 1987, A comparative 1H-n.m.r. study of Mur and Ac-L-Ala-D-iGln (MDP) and its analogue murabutide: evidence for a structure involving two successive beta-turns in MDP, Carbohydr. Res., 16:23.

Finkelman, F. D., Holmes, J., Katona, I. M., Urban, J. F. Jr., Beckmann, M. P., Park, L. S., Schooley, K. A., Coffman, R. L., Mosmann, T. R., and Paul, W. E., 1990, Lymphokine control of in vivo immunoglobulin isotype selection, Annu. Rev. Immunol., 8:303.

Fraser-Smith, E. B., Waters, R. V. and Matthews, T. R., 1982, Correlation between in vivo anti-Pseudomonas and anti-Candida activities and clearance of carbon by the reticuloendothelial system for various muramyl dipeptide analogs, using normal and immunosuppressed mice, Infect. Immun., 35:105.

Geczy, A. F., Alexander, K., Bashir, H. V., Edmonds, J. P., Upfold, L., and Sullivan, J., 1983, HLA-B27, Klebsiella and ankylosing spondylitis: biological and chemical studies, Immun. Rev., 70:23.

Girard, M., Kieny, M.-P. and Pinter, A., 1991, Immunization of chimpanzees confers protection against challenge with human immunodeficiency virus, Proc. Natl. Acad. Sci. USA, 88:542.

Gosselin, R. E., Hodge, H. C., Smith, R. P. and Gleason, M. N., 1976, "Clinical toxicology of commercial products," Fourth Edition, Williams & Wilkins Company, Baltimore.

Hayward, A. R., Burger, R., Scheper, R. and Arvin, A. M., 1991, Major histocompatibility complex restriction of T-cell responses to varicella-zoster virus in guinea pigs, J. Virol., 65:1491.

Hunter, R. L., Strickland, F. and Kezdy, F., 1981, The adjuvant activity of nonionic block polymer surfactants. I. The role of hydrophile-lipophile balance, J. Immun., 127:1244.

Jones, W. R., Bradley, J., Judd, S. J., Denholm, E. H., Ing, R. M. Y., Mueller, U. W., Powell, J., Griffin, P. W. and Stevens, V. C., 1988, Phase I clinical trials of a World Health organization birth control vaccine, Lancet, i:1295.

Kaminski, M. S., Kitamura, K., Maloney, D. G., Campbell, M. J. and Levy, R., 1986, Importance of antibody isotype in monoclonal anti-idiotype therapy of murine B cell lymphoma. A study of hybridoma class-switch variants, J. Immunol., 136:1123.

Kenney, J. S., Hughes, B. W., Masada, M. P. and Allison, A. C., 1989, Influence of adjuvants on the quantity, affinity, isotype and epitope specificity of murine antibodies, J. Immun. Methods, 121:157.

MacLennan, I., 1991, The centre of hypermutation, Nature, 354:352.

Marx, P. A., Pedersen, N. C., Lerche, N. W., Osborn, K. G., Lowenstine, L. J., Lackner, A. A., Maul, D. H., Kwang, H. -S., Kluge, J. D., Zaiss, C. P., Sharpe, V., Spinner, A. P., Allison, A. C. and Gardner, M. B., 1986, Prevention of simian acquired immunodeficiency syndrome with a formalin-inactivated Type D retrovirus vaccine, J. Virol., 60:431.

Metlay, J. P., Pure, E. and Steinman, R. M., 1990, Control of immune response at the level of antigen-presenting cells: a comparison of the function of dendritic cells and B-lymphocyctes, Adv. Immunol., 47:45.

Morgan, A. J., Allison, A. C., Finerty, S., Scullion, F. T., Byars, N. E., and Epstein, M. A., 1989, Validation of a first generation Epstein-Barr virus vaccine preparation suitable for human use, J. Med. Virol., 29:74.

Morris, A. G., Lin, Y. -L. and Askonas, B. A., 1982, Immune interferon release when a cloned cytotoxic T-cell line meets its correct influenza-infected target, Nature, 295:150.

Mosmann, T. R. and Coffman, R. L., 1989, TH1 and TH2 cells: different patterns of lymphokine secretion lead to different functional properties, Annu. Rev. Immunol., 7:145.

Murray, R., Cohen, P. and Hardegree, M. C., 1972, Mineral oil adjuvants: biological and chemical studies, Ann. Allergy, 30:146.

Murphey-Corb, M., Martin, L. N., Davison-Fairburn, B., Okawa, S., Baskin, G. B., Zhang, J. -Y., Montelaro, R. C., Miller, M., West, M., Allison, A. C. and Eppstein, D. A., 1989, A formalin inactivated whole simian immunodeficiency virus vaccine confers protection in macaques, Science, 246:1293.

Nagao, S. and Tanaka, A., 1980, Muramyl dipeptide-induced adjuvant arthritis, Infect. Immun., 28:624.

Nicholson, K. G., Tyrrell, D. A. J., Harrison, P., Potter, C. W., Jennings, R., Clark, A., Schild, G. C., Wood, J. M., Yells, R., Seagrott, V., Huggens, A. and Anderson, S. G., 1979, Clinical studies of monovalent inactivated whole virus and subunit A/USSR/77 (H_1N_1) vaccine: serological and clinical reactions, J. Biol. Stand., 7:123.

Potter, M. and Boyce, C. R., 1972, Induction of plasma cell neoplasms in BALB/c strain mice with mineral oil and mineral oil adjuvants, Nature, 193:1086.

Reichmann, L., Clark, M., Waldmann, H. and Winter, G., 1988, Reshaping human antibodies for therapy, Nature, 332:323.

Ribi, E., Ulrich, J. T. and Masihi, K. N., 1987, Immunopotentiating activities of monophosphoryl lipid A, in: "Immunopharmacology of Infectious Diseases: Vaccine Adjuvants and Modulators of Non-Specific Resistance," J. A. Majde, ed., Alan R. Liss, New York.

Rickinson, A. B., Rowe, M., Hart, I. J., Yao, W. Y., Henderson, L. E., Robin, H. and Epstein, M. A., 1984, T cell-mediated regression of "spontaneous" and of Epstein-Barr virus-induced B-cell transformation in vitro: studies with cyclosporin A, Cell. Immunol, 87:646.

Steplewski, Z., Spira, G., Blasczyc, M., Lusbeck, M. D., Radlmuch, A., Illges, H., Herlyn, D., Rajewsky, K. and Scharff, M., 1985, Isolation and characterization of anti-monosialoganglioside monoclonal antibody 19-9S class switch variants, Proc. Natl, Acad, Sci. USA, 82:3653.

Takahashi, H., Takeshita, T., Morein, B., Putney, S., Germain, R. N. and Berzofsky, J. A., 1990, Induction of CD8[+] cytotoxic T-cells by immunization with purified HIV-1 envelope protein in ISCOMS, Nature, 344:873.

Tew, J. G., Kosco, M. H. and Szakal, A. K., 1989, The alternative antigen pathway, Immunol. Today, 10:229.

Unkeless, J. C. Scaling, E. and Freedman, V. H., 1988, Structure and function of human and murine receptors for IgG, Ann. Rev. Immun., 6:251.

ORAL ADMINISTRATION OF PEPTIDES: BYPASSING A HOSTILE MILIEU

Murray Saffran

Department of Biochemistry and Molecular Biology
Medical College of Ohio
Toledo, Ohio 43699, USA

INTRODUCTION

As a peptide drug passes through the alimentary canal it runs the gauntlet of hostile environments. These include the very acid pH of the stomach, the secreted proteolytic enzymes in the stomach and small intestine, the cell-membrane-bound enzymes of the brush border of the small intestine, and the host of bacterial enzymes that may be liberated in the colon. Of these, the proteolytic enzymes of pancreatic origin in the small intestine, trypsin and chymotrypsin, are the most powerful destructive agents.

PROTECTION IN THE STOMACH

Peptide drugs can be protected against acid and enzymes by coating the drug with an impervious barrier, thereby preventing contact of the drug with the destructive agent. For protection against the acid and enzymes in the stomach, the peptide drug is coated with an enteric substance, usually an organic acid, which is insoluble in water at the acid pH of the stomach. This is a time-honoured strategy for the oral administration of acid-labile substances and is still an excellent approach. The enteric coating becomes water-soluble when the dosage form reaches the small intestine, where the pH is close to neutrality. The coating dissolves and exposes the drug to the environment of the small intestine. Now the problem is to protect the peptide drug from hydrolysis by enzymes in the small intestine.

PROTECTION IN THE SMALL INTESTINE

Emulsions and Micelles

The peptide drug can be absorbed quickly before it can be destroyed. This is usually attempted by disguising the drug as a lipid; lipids are not only protected in the lumen of the small intestine by the formation of micelles with the aid of bile sales, excluding much of the water-borne enzymatic activity, but the lipid micelles tend to be absorbed quickly via lacteals, vessels that drain lymph from the intestine into the

Targeting of Drugs 3: The Challenge of Peptides and Proteins
Edited by G. Gregoriadis et al., Plenum Press, New York, 1992

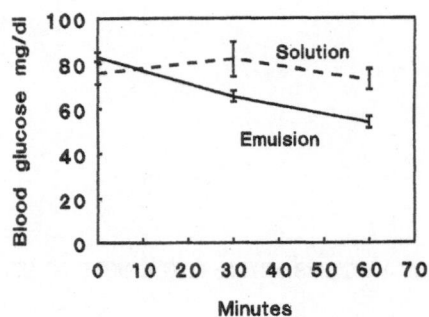

Fig. 1. Blood glucose in normal rats after direct administration
into the duodenum of a solution (dashed line) or an
emulsion (solid line) of insulin. After data in Engel,
Riggi and Fahrenbach, 1968.

peripheral circulation via the thoracic duct. If the disguise is
successful, some of the drug can win the race against destruction and
survive.

In an early study by Murlin et al (1940) insulin was dissolved in a
solution of hexylresorcinol, a lipid-like compound that has detergent
properties, and which was shown to promote the absorption of insulin from
the gastrointestinal tract in animals. The solution was administered to
fasting patients with diabetes mellitus. In some of the subjects there
were decreases in both the glucose excreted in the urine and in blood
levels of glucose. However, because all subjects were also given
subcutaneous doses of insulin at the same time, the effects of the oral
insulin could only be estimated.

The simplest method of disguising the peptides as a lipid involves
the formation of an emulsion of an aqueous solution of the drug with a
lipid or lipid mixture. Such emulsions were initially administered to
experimental animals by direct instillation into segments of the
gastrointestinal tract. Engel, Riggi and Fahrenbach (1968) emulsified an
aqueous solution of insulin and zinc chloride with palmitic acid in
trioctanoin. This emulsion was then reemulsified with an aqueous
solution of the detergent, sodium lauryl sulphate (sodium dodecyl
sulphate), to yield a stable water-in-oil-in-water emulsion. The emulsion
was injected directly into the duodenum of anaesthetized normal rats or
gerbils and blood glucose levels were measured at 30 and 60 minutes. The
results in rats are illustrated in Fig. 1, which compares the effect of the
insulin emulsion with an equivalent dose of insulin in solution. The
insulin emulsion lowered the glucose level significantly as early as 30
minutes after administration, while the insulin solution was relatively
ineffective. Paradoxically, similar insulin emulsions were ineffective in
alloxan-diabetic rats. This was blamed on the altered gastrointestinal
absorption in alloxan diabetes. Shichiri and his colleagues (Shichiri et
al, 1974) repeated the work of Engel et al in rabbits, with direct instillation of the emulsions into the jejunum. They measured plasma insulin in
addition to blood glucose. Peaks of plasma insulin occurred at 30 minutes,
while a nadir in blood glucose occurred at 90 minutes (Fig. 2). In both
reports the responses were dose-related. Shichiri et al (1974) further
exposed the insulin emulsions to proteolytic enzymes and demonstrated that
the insulin was protected against hydrolysis.

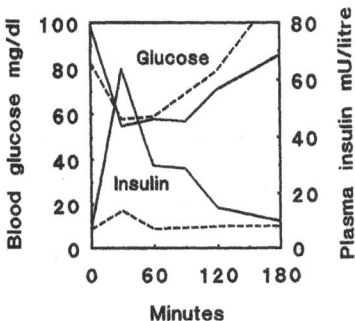

Fig. 2. Blood glucose and plasma insulin after direct instillation
of an emulsion (solid line) or a solution (dashed line) of
insulin into the jejunum of anaesthetised normal rabbits.
From data of Shichiri et al, 1974.

Another approach was used by Teng (1986). A solution of insulin was
mixed with a solution of sodium dodecyl sulphate (SDS). SDS forms a
complex with insulin in a ratio of about 1.4:1, which is resistant to
attack by stomach acid. The complex is further combined with a solution
of a quaternary amine, such as triethylamine hydrochloride. The so-
called "sandwich" not only protects the insulin against hydrolytic
damage, but also facilitates absorption via the lipid route. Such
insulin complexes given orally to streptozotocin-treated rats resulted in
a statistically significant decrease in blood glucose 2 and 3 hours after
administration, with a return toward control values at 5 hours. The mean
decrease at the nadir was 119 mg per deciliter. However, the mean blood
glucose level of the rats, before insulin was given, was not in the
diabetic range. Therefore, there is need to repeat the experiment on
proven diabetic rats. In non diabetic rats, the oral insulin complex was
followed by a maximum decrease of 35 mg per deciliter 1 hour later. The
decrease was not statistically significant. Plasma insulin levels were
not measured.

More recently an insulin-lipid emulsion was administered orally to
human subjects with insulin-dependent diabetes mellitus (Cho and Flynn,
1989). The insulin was emulsified with a mixture containing lecithin,
fatty acids and cholesterol. Peaks of insulin were measured in plasma
between 0.5 and 2 hours. Unfortunately the human trials were flawed and
must be repeated (Stinson, 1991).

Liposomes

Instead of a lipid disguise for the peptide drug itself, the peptide
solution can be entrapped into liposomes, spheres of phospholipids. So
many attempts have been made to deliver liposomes containing peptide
drugs orally that a special chapter devoted to liposomes would be
appropriate. In outline, an aqueous solution of the drug is mixed with a
solution of phospholipid, or is exposed to a layer of phospholipid on the
surface of a vessel. Because phospholipid molecules have a hydrophilic
head of phosphate and a long hydrophobic tail of fatty acids, the peptide
drug solution is attracted to the phosphate. Physical forces then shape
the phospholipids into spheres, with the lipid on the outer surface, and
the hydrophilic groups, now including the drug, on the inside of the
spheres. The whole assembly has the properties of a lipid, and may
theoretically be absorbed by the lipid transporting system of the
gastrointestinal tract. In early trials, insulin-containing liposomes

were given by stomach tube to normal rats and to rats made diabetic with streptozotocin (Dapergolas and Gregoriadis, 1976; Ryman and Patel, 1976). Both groups observed a fall in the level of blood glucose after the insulin-containing liposomes, but only Dapergolas and Gregoriadis reported a fall in normal rats. The literature contains about an equal number of successes and failures with oral insulin in liposomes; consensus must await more rigorous trials.

Polymeric nanospheres

An insulin solution can be entrapped in the matrix of a polymer fashioned into spheres of microscopic size, 50 to 100 nanometers in diameter (Damge et al, 1988). Such tiny spheres are apparently phagocytized by lymphoid tissue in the intestinal wall and are sent via the lymph vessels into the circulation. The uptake of nanospheres is rapid, before the drug in them can be digested. The drug-filled nanospheres are trapped in capillary beds in organs such as the spleen (Jani et al, 1990). Because the polymer slowly hydrolyses, the drug is released slowly over a long time.

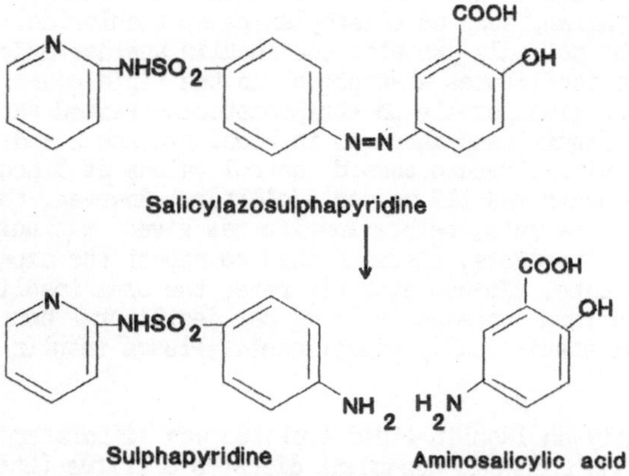

Fig. 3.   Reductive cleavage of salicylazosulphapyridine by colonic bacteria.

Inhibiting Proteolysis

Another approach is to inhibit temporarily the proteolytic action of the enzymes in the small intestine by the prior or simultaneous feeding of enzyme inhibitors. Inhibitors range from diisopropylfluorophosphate (DFP), a very toxic compound, to trypsin inhibitors derived from natural sources, such as aprotinin, from bovine lung, and a trypsin inhibitor from soy beans. Aprotinin has the advantage of inhibiting both trypsin and chymotrypsin. In most trials, the oral activity of a peptide drug is enhanced by the inhibitor (Laskowski et al, 1958; Saffran et al, 1988).

A combination of lipid disguise and enzyme inhibition has recently been applied to the oral administration of insulin in man (Cho and Flynn, 1989).

84

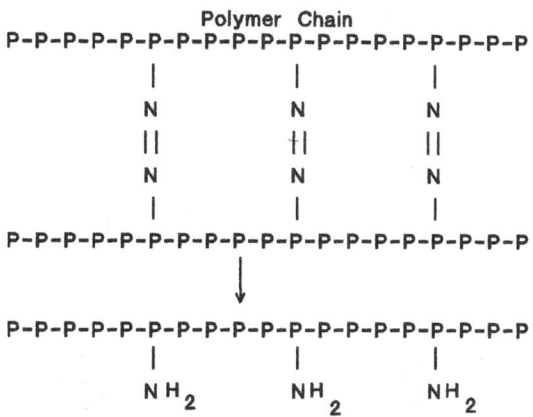

Fig. 4.   Reduction of azo crosslinks in an azopolymer.

DELIVERY TO THE COLON

Yet another approach takes advantage of the relatively benign
environment of the colon for the exposure of the peptide drug to an
absorbing surface. Although the colon is not as efficient as the small
intestine at absorption, the absence of trypsin and chymotrypsin makes it
a very attractive target for delivery of a peptide drug.

One or Two Coats

Oral delivery to the colon is possible in several ways. The pH of
the human distal ileum and colon is said to be just above 7. Capsules
containing a drug can be coated with an acrylic polymer that only
dissolves at pH >7. The coating protects the capsule and its contents
during the journey through the stomach and small intestine and delivers
the load at the junction between the small and large intestine (Dew et
al, 1982).

Transit time from the mouth to the colon is about 4 to 5 hours in
human subjects. Simple coating of the drug with a protective substance
that erodes completely in 4 to 6 hours would deliver the drug to the colon
some of the time. A portion of the dosage forms would expose the drug
prematurely. Late exposure of the drug far into the colon poses another
danger in the form of trapping of the drug in the midst of solid feces,
which might erect a solid barrier between the drug and the colonic
epithelium to prevent or at least delay absorption. Because the stomach
emptying time is the most variable component of the mouth to colon transit
time, it is possible to coat the peptide drug with two layers: the outer
layer is an enteric coating, which would dissolve away when the dosage form
enters the small intestine from the stomach; the inner coating is designed
to erode away in the 4 hours transit time from stomach to colon. In this
case the timing signal is given by the exit from the stomach.

Harnessing Colonic Bacteria

Another strategy uses the colon itself for the timing mechanism.
The colon is the residence of billions of microorganisms, with metabolic

spectra very different from the host's. Enzymatic activity specific to the bacteria can be used to trigger the delivery of the drug when the dosage form enters the colon. One of these is bacterial α-glycosidase (Friend and Chang, 1984). Coupling a drug to a polysaccharide by an - glycosidic linkage can deliver corticoids to the colon, where bacterial action cleaves the drug from a polysaccharide that prevents absorption in the small intestine. This delivery system leaves the drug pendant to the delivery molecule, and thus open to degradation during passage. Therefore it cannot be used for digestible drugs like peptides. However, if the polysaccharide is fashioned into a closed covering, then the load can be protected against digestion during passage and freed by bacterial action in the colon.

Other chemical groups susceptible to bacterial attack can be harnessed for drug delivery. The drug, salicylazosulphapyridine, consisting of salicylic acid and sulphapyridine joined via an azo bond, is used in the treatment of inflammatory colon disease. The complex is insoluble and, after oral administration, passes unabsorbed through the small intestine. In the colon, however, the azo bond is reduced to two amines by bacteria and the two parts of the drug separate (Fig. 3). The salicylate half is anti-inflammatory, while the sulphapyridine carrier is absorbed and detoxified in the liver (Peppercorn and Goldman, 1972).

A polymer can be designed with cross-linkages containing azo bonds. The cross-linked polymer is relatively waterproof and can serve as a protective coating in the stomach and small intestine for oral administration of a peptide drug. However, when the dosage form reaches the colon, the azo bonds are cleaved (Fig. 4), and the coating becomes permeable to water, thereby delivering the load in the colon (Saffran et al, 1986). The colonic trigger is theoretically more reliable than a gastric (stomach) trigger because it is activated at the delivery site instead of several metres upstream in the alimentary canal.

Protection of the peptide drug against hazards is only half the battle. Once delivered safely to a desired site in the intestine, the peptide drug, usually of larger size than most orally administered drugs, must traverse multiple membranes before entering the bloodstream. Progress is being made in facilitation of passage of peptide drugs from the intestine to the blood and will probably be the subject of a future report in this series.

SUMMARY

The oral administration of peptide drugs faces two formidable problems. The first is protection against the hazards of digestion in the stomach and small intestine. The second is absorption from the gastrointestinal tract in the absence of a carrier system for peptides of more than three amino acid residues. By disguising the peptide drug as a lipid, the absorption of the drug can be so fast that much of it will survive digestion. Alternatively, destruction by digestive peptidases can be prevented by the co-administration of inhibitors of peptidases. The drug can also be protected by coating of the dosage form with impervious polymers that are designed to deliver the drug in the colon, which is devoid of secreted digestive enzymes.

Acknowledgements

The development and synthesis of the azopolymer was carried out by Drs. C. Savariar, G.S. Kumar and D.C. Neckers of the Department of Chemistry, Bowling Green State University, Bowling Green, Ohio, with support from the Eli Lilly company to D.C.N. and a grant from the National Institutes of Health to M.S.

REFERENCES

Cho, Y. W., and Flynn, M., 1989, Oral delivery of insulin, The Lancet, 2:1518.

Damgé, C., Michel, C., Aprahamian, M., and Couvreur, P., 1988, New approach for oral administration of insulin with polyalkylacrylate nanocapsules as drug carrier, Diabetes 37:246.

Dapergolas, G., and Gregoriadis, G., 1976, Hypoglycaemic effect of liposome-entrapped insulin administered intragastrically into rats, Lancet, 2: 824.

Dew, M.J., Hughes, P.J., Lee, M.G., Evans, B.K., and Rhodes, J., 1982, An oral preparation to release drugs in the human colon, Br. J. Clin. Pharmacol, 14:405.

Engel, R.H., Riggi, S.J., and Fahrenbach, H.J., 1968, Insulin: intestinal absorption as water-in-oil-in-water emulsions, Nature, 219:856.

Jani, P., Halbert, G.W., Langbridge, J., and Florence, A.T., 1990, Nanoparticle uptake by the rat gastrointestinal mucosa: Quantitation and particle size dependency, J. Pharm. Pharmacol, 42:821.

Laskowski, M., Jr., Haessler, H.A., Miech, R.P., Peanasky, R.J. and Laskowski, M., 1958, Effect of trypsin inhibitor on passage of insulin across the intestinal barrier, Science, 127:1115.

Peppercorn, M.A., and Goldman, P, 1974, The role of intestinal bacteria in the metabolism of salicylazosulphapyridine, J. Pharmac. Exper. Therap., 181:555.

Murlin, J.R., Gibbs, C.B.F., Romansky, M.J., Steinhausen, T.B., and Truax, F.L., 1940, Effectiveness of per-oral insulin in human diabetes, J. Clin. Invest., 19:709.

Patel, H.M., and Ryman, B.E., 1976, Oral administration of insulin by encapsulation within liposomes, FEBS Lett., 62:60.

Saffran, M., Bedra, C., Kumar, G.S., and Neckers, D.C., 1988, Vasopressin: A model for the study of effects of additives on the oral and rectal administration of peptide drugs, J. Pharmaceut. Sci., 77:33.

Saffran, M., Kumar, G.S., Savariar, C., Burnham, J.C., Williams, F., and Neckers, D.C., 1986, A new approach to the oral administration of insulin and other peptide drugs, Science, 233:1081.

Shichiri, M., Shimizu, Y., Yoshida, Y., Kawamori, R., Fukuchi, M., Shigeta, Y. and Abe, H., 1974, Enteral absorption of water-in-oil-in-water insulin emulsions in rabbits, Diabetologia, 10:317.

Stinson, S.C., 1991, New drugs under development for diabetes, Chemical & Engineering News, 69:35.

Teng, L.-N. L., 1986, Orally administered biologically active peptides and proteins, United States Patent, 4,592,820.

ORAL ADMINISTRATION OF INSULIN: IMITATING THE NATURAL PATHWAY

Murray Saffran

Department of Biochemistry and Molecular Biology
Medical College of Ohio
Toledo, Ohio 43699, USA

INTRODUCTION

Before the discovery of insulin by Banting and Best in 1921, a
diagnosis of diabetes, especially in a young child, was a sentence of
death. Since then, insulin has given the patient with diabetes mellitus
many years of life at the expense of daily injections. So far no
practical alternative has been developed to the injection route for
administration of insulin. As time passed and the initial gratitude for
the life-preserving effect of insulin became commonplace, the drawbacks
of insulin injections became apparent. First of all there is the
discomfort of the injection, now lessened by the development of thinner
needles and needle lubricants, the danger of infection, mitigated by
convenient alcohol swabs and the disposable needle and syringe, the
desire for privacy, decreased somewhat by the development of injection
pens and external pumps, and the bother of carrying around and caring for
supplies of insulin solutions. With prolonged life came the dilemma of
running out of suitable injection sites, as old ones became pin cushioned
and less receptive to the needle. But more important than all of these
drawbacks of injected insulin, the insulin is delivered to the wrong
place.

PHYSIOLOGICAL DELIVERY OF INSULIN

Physiological insulin, made in the islets of the pancreas, is
secreted into blood vessels that lead directly into the hepatic portal
vein, the same vein that carries the products of digestion from the
intestine to the liver. Because insulin's task is to control the fate of
the absorbed foods in the liver, the raw material and the controlling
agent are delivered together to the organ in which further processing
occurs. In the liver, insulin's primary actions are (a) to suppress the
liver's ability to make new glucose out of non-glucose raw material,
mostly amino acids and lactic acid, and (b) to stimulate the storage of
glucose arriving from the intestine as glycogen. The liver is very
sensitive to low concentrations of insulin arriving via the hepatic
portal vein. Only small amounts of insulin are needed to keep the liver
under metabolic control (Fig. 1) (Rizza et al, 1981). The liver, in
turn, can control the amount of insulin that passes through it (Ishida et
al, 1984). If a surge of insulin reaches the liver from the pancreas,

Targeting of Drugs 3: The Challenge of Peptides and Proteins
Edited by G. Gregoriadis et al., Plenum Press, New York, 1992

89

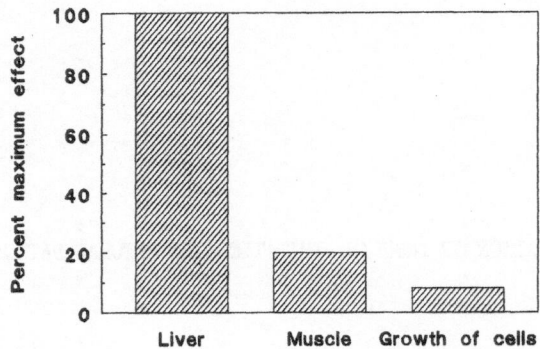

Fig. 1. The percent of maximal response elicited by a concentration of 50 mUnits/litre on glucose production by the liver, glucose utilisation by muscle and on cell division in tissue culture. From data in Rizza, Mandarino, and Gerich, 1981, and Stout, 1991.

the liver can inactivate the excess and keep the rest of the body from being exposed to too much insulin.

DELIVERY OF INJECTED INSULIN

The pancreas of the patient with insulin-dependent diabetes mellitus has lost the ability to make and secrete insulin. Instead the insulin is injected under the skin in the leg, the abdomen or the arm. The insulin leaves the site of injection and is drained into tissue capillaries and lymph channels that lead it to the venous return to the heart. The heart

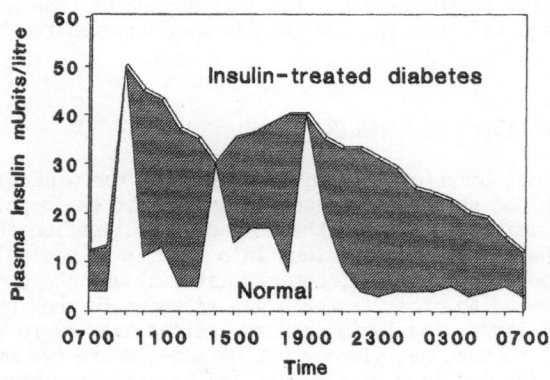

Fig. 2. A comparison of peripheral plasma levels of insulin in a normal person and in a diabetic patient treated with one injection of long-lasting insulin in the morning, plus injections of short acting insulin just before meals. The aim of the insulin therapy is to match the normal peaks that occur after meals. From data in Schade et al, 1983).

then pumps the insulin-containing blood all over the body. The liver receives a fraction of the blood containing insulin and responds to it. However, the amount of insulin that bathes the other organs is no longer under the control of the liver. During part of the day and night, the tissues of the insulin-injected individual may be exposed to higher-than-normal concentrations of insulin (Fig. 2). Much of the action of insulin after injection is exerted on the muscle cells, which respond by opening channels for glucose into the cells, which then use the glucose for energy purposes.

Hypoglycaemia

Glucose is virtually the only energy source of the brain. If glucose levels in the blood fall too low, the brain is starved of its energy source and the person suffers from restricted brain function. Under extreme glucose deprivation there can be slowing of mental processes, coma, and even death. Luckily, the counter-hormones prevent such extreme effects of insulin from the pancreas. The effect of insulin in the liver is limited by the actions of counter-hormones, such as adrenalin, glucagon, growth hormone and the corticosteroids. As a result, the action of insulin on the liver is restrained and seldom goes too far. However, injected insulin, acting mainly on muscle cells, has almost unrestricted effects on the muscles. Of the counter-hormones, only adrenalin can quickly offset the response to excessive insulin. It is not uncommon for the diabetic patient to experience the effects of excessive insulin after injection if unforeseen circumstances occur, such as a missed meal or vigorous exercise.

Insulin and Growth Factors

Insulin belongs to a family of peptide agents that include the insulin-like growth factors, IGF-I and IGF-II. The IGFs stimulate cell division and protein synthesis. Although insulin has little IGF-like activity at low concentrations, at concentrations over 100 mUnits/litre, which can be reached after insulin injection, the ability of insulin to stimulate cell proliferation increases significantly (Fig. 1) (Stout, 1991). If these effects occur within blood vessels, this leads to local

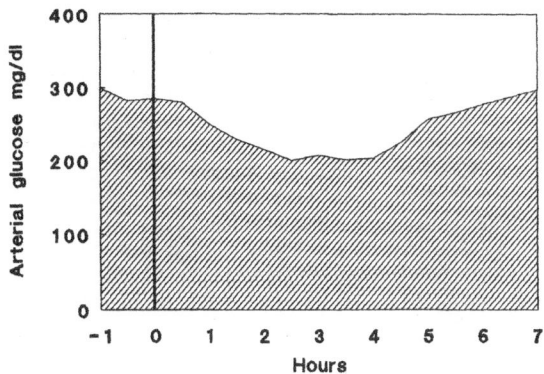

Fig. 3.　Effect on arterial plasma glucose of a single oral dose of insulin, given at 0 hour in an azopolymer-coated capsule to a fasting diabetic dog.the enteral delivery of insulin (Gwinup and Elias, 1990).

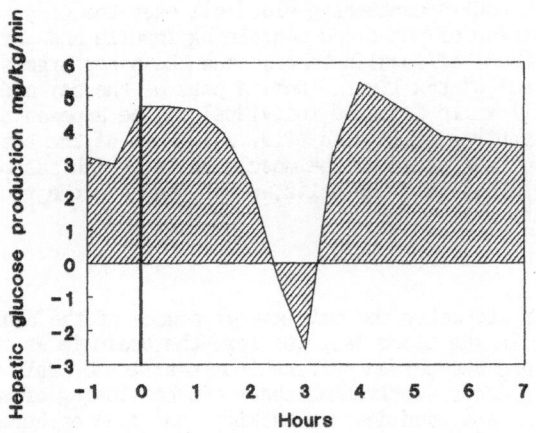

Fig. 4. Effect on hepatic glucose production of a single oral dose
of insulin, given at 0 hour in an azopolymer-coated capsule
to the same fasting diabetic dog as in Fig. 3.

thickening of the wall, increased turbulence of blood flow, and
eventually to the formation of an atherosclerotic site or narrowing of
the vessel (Stolar, 1988). If this occurs in the heart, a coronary
vessel can be involved, leading to a heart attack. There is growing
concern that long-term injections with insulin may contribute to the
increased incidence of heart disease among diabetic patients (Saffran,
1989).

For many reasons it would be desirable to substitute for peripheral
injection of insulin a method for delivery of insulin into the hepatic
portal vein, mimicking nature's route of administration,. Because much
of the alimentary canal drains into the portal vein, an intestinal
delivery system would be best. Therefore there is renewed interest in the
enteral delivery of insulin (Gwinup and Elias, 1990).

Oral Delivery of One Dose of Insulin

Using insulin in gelatin capsules coated with a polymer cross-linked
with an azo group, we administered insulin orally to diabetic dogs
(Saffran et al, 1990). We found, like many others before us, that a
single dose of insulin did little to normalize the very high levels of
blood glucose so characteristic of diabetes (Fig. 3). We realized that a
single oral dose should not do very much, because most of the action of
the oral insulin was exerted in the liver on the supply of glucose to the
blood. The single dose shuts off the source of glucose only for a short
time (Fig. 4). During that time little insulin gets through the liver to
stimulate the use of glucose by the muscle (Fig. 5). The small decline
in blood glucose is due almost entirely to the use of glucose by the
brain, which does not need insulin to utilise glucose. Because the use
of glucose by the brain is limited, the fall in blood glucose after a
single dose of insulin is also limited.

Oral Delivery of Multiple Doses of Insulin

If, however, the oral dose of insulin is repeated often enough to
suppress glucose formation by the liver for a long time, then utilisation
of glucose by the brain would eventually lower the blood level of glucose

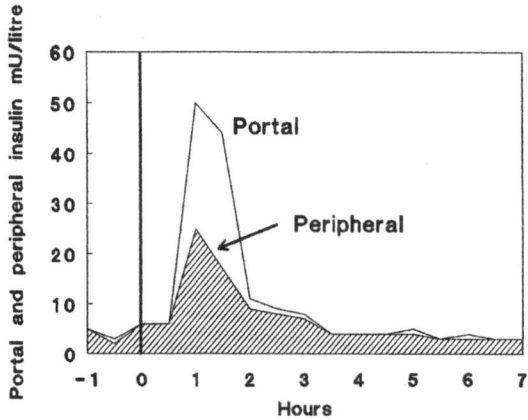

Fig. 5. Effect on plasma insulin in blood from a peripheral artery
and from the portal vein of a single oral dose of insulin,
given at 0 hour in an azopolymer-coated capsule to the same
fasting diabetic dog as in Fig. 3.

to normal limits. Therefore, we gave repeated oral doses of insulin to
diabetic dogs (Fig. 6) and observed a steady fall in blood glucose toward
normal (Fig. 7). That this was due to an effect on the liver was
confirmed by direct measurement of the amount of glucose secreted into
the blood by the liver. During the period of repeated oral insulin,
glucose production by the liver was almost completely suppressed (Fig.
8).

Oral insulin, then, imitates the natural route of delivery of
insulin from the pancreas and repeated doses may be able to normalize the
blood glucose levels without subjecting the rest of the body to
abnormally high concentrations of insulin. Will the control of diabetes
with oral insulin decrease the cardiovascular complications in diabetes?

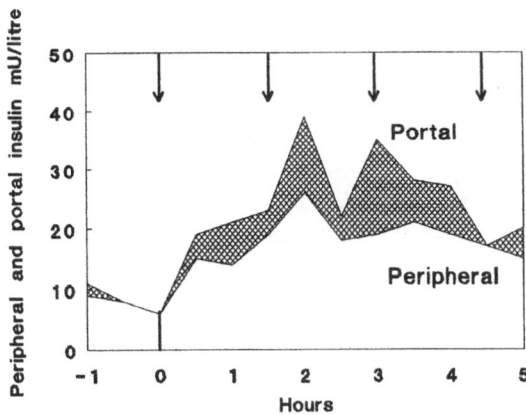

Fig. 6. Plasma insulin in a fasting diabetic dog given repeated
oral doses of insulin at the arrows

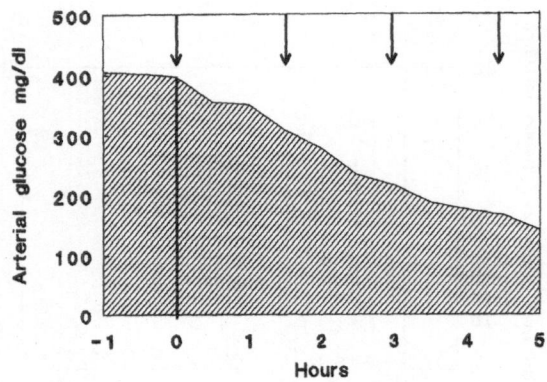

Fig. 7. Arterial plasma glucose in the same dog as in Fig. 6. is

That remains to be seen in the future. As the oral delivery of peptides is developed, consideration must be given to the possibility that more is changed than merely the route of administration; the effect of the liver on the drug and of the drug on the liver must be assessed.

CONCLUSIONS

Oral administration of insulin delivers it by a more physiologic route than subcutaneous injection. Multiple oral doses are required to bring the blood glucose levels down into the normal range. Oral insulin may decrease the incidence of hyperglycaemia reactions and may also decrease the incidence of some complications of diabetes mellitus.

Acknowledgements

The azopolymer used for oral delivery of insulin was supplied by Drs. D.C. Neckers and G.S. Kumar, of the Department of Chemistry, Bowling

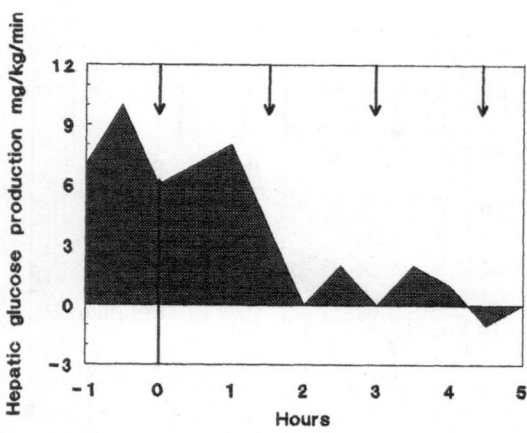

Fig. 8. Hepatic glucose production in the same dog as in Fig. 6.

Green State University, Bowling Green, Ohio, USA. Experiments on diabetic dogs were carried out with James Pena and Drs. J.B. Field, R. H. Jones and Y. Okuda in the Department of Medicine, Baylor College of Medicine, Houston, Texas, USA. The work was supported by grants from the National Institutes of Health to M.S. and J.B. Field, Biomedical Research Support, to M.S., Eli Lilly Company, to D.C. Neckers, and the Diabetes Research and Education Foundation, to M.S.

REFERENCES

Gwinup, G., and Elias, A. N., 1990, The physiologic replacement of insulin, New Eng. J. Med., 322:333.

Ishida, T., Chap, Z., Chou, J., Lewis, R. M., Hartley, C. J., Entman, M.L., and Field, J. B., 1984, Hepatic extraction of exogenous insulin in depancreatized conscious dogs, Am. J. Physiol, 246:E369.

Rizza, R. A., Mandarino, L. J., and Gerich, J. R., 1981, Dose-response characteristics for effects of insulin on production and utilization of glucose in man, Am. J. Physiol, 235: E630.

Saffran, M., 1989, Is insulin a factor in the genesis of the vascular complications of diabetes? Trends Endocrinol. Metab., 1:56.

Saffran, M., Kumar, G. S., Neckers, D. C., Pena, J., Jones, R. H., and Field, J. B., 1990, Biodegradable azopolymer coating for oral delivery of peptide drugs, Biochem. Soc. Trans., 18:752.

Schade, D. S., Santiago, J. V., Skyler, J. S., and Rizza, R. A., 1983, "Intensive Insulin Therapy", Medical Examination Publishing Company (Exerpta Medica), Princeton.

Stolar, M. W., 1988, Atherosclerosis in diabetes: The role of hyperinsulinemia, Metabolism, 37 (suppl 1):1.

Stout, R. W., 1991, Insulin as a mitogenic factor: Role in the pathogenesis of cardiovascular disease, Am. J. Med., 90 (suppl 2A): 62S and 65S.

NEUROPEPTIDE-MEDIATED GROWTH OF NORMAL AND CANCER CELLS:

INHIBITION BY BROAD SPECTRUM ANTAGONISTS

Enrique Rozengurt and Tariq Sethi

Imperial Cancer Research Fund
PO Box 123, Lincoln's Inn Fields
London WC2A 3PX, U.K.

INTRODUCTION

Cancer cells are characterized by unrestrained growth. Until recently pharmacologists have concentrated on disrupting cell proliferation to kill cancer cells. The cytotoxic drugs developed have been extremely effective in the minority of rapidly growing tumours, but their effects on normal dividing tissues have led to dose-limiting toxicities. Attention has now moved to the biology of the cancer cell, in the expectation that elucidation of the factors stimulating tumour growth and their modes of action will permit rational development of novel and specific antitumour agents, with activity in a wider range of solid tumours.

In recent years it has become evident that neoplastic cells acquire complete or partial independence of growth control through different mechanisms (Rozengurt, 1983; Sporn and Roberts, 1985; Goustin et al., 1986; Cross & Dexter, 1991). These include production of autocrine or paracrine growth factors, alterations in the number or structure of cellular receptors and changes in the activity of post-receptor signalling pathways (Sager, 1989; Bishop, 1991). Cancers are thought to result from the accumulation of multiple genetic changes, causing activation of oncogenes and deletion of tumour-suppressor genes. The discovery that many oncogenes code for growth factors, their receptors or for proteins involved in intracellular signalling has been central to this thesis.

Because of the complex interactions between growth factors in vivo, direct evidence for their effects has depended upon the development of homogeneous cell lines in vitro. The non-tumorigenic murine Swiss 3T3 fibroblast line has proved useful for identifying both the extracellular factors that modulate cell growth and the early signals and molecular events that lead to mitogenesis. These cells cease to proliferate when the medium is depleted of its growth-promoting activity and can be stimulated to re-initiate DNA synthesis and cell division either by replenishing the medium with fresh serum, or by the addition of purified growth factors, pharmacological agents or a variety of neuropeptides (Rozengurt, 1985). Studies done using such quiescent cells and defined combinations of growth factors have revealed the existence of potent and

Targeting of Drugs 3: The Challenge of Peptides and Proteins
Edited by G. Gregoriadis et al., Plenum Press, New York, 1992

Table 1. Events in the action of bombesin in Swiss 3T3 cells

Event	Reference
Binding to specific receptors	Zachary & Rozengurt (1985a)
Cross-linking to M_r 75,000–85,000 glycoprotein	Zachary & Rozengurt (1987a); Kris et al (1987); Sinnett-Smith et al (1988)
Ligand internalization and degradation	Zachary & Rozengurt (1987b)
Down-regulation of receptors	Millar & Rozengurt (1990)
Activation of PKC (intact cells)	Zachary et al (1986)
Activation of PKC (permeabilized cells)	Erusalimsky et al (1988)
Elevation of DAG levels	Issandou and Rozengurt, 1990
Ins(1,4,5)P$_3$ production	Lopez-Rivas et al (1987) Nanberg & Rozengurt (1988)
Ca^{2+} mobilization	Mendoza et al (1986); Lopez-Rivas et al (1987)
Tyrosine phosphorylation	Zachary et al (1991)
Na$^+$ influx and Na$^+$/K$^+$ pump	Mendoza et al (1986)
Transmodulation of EGF receptor	Zachery et al (1986)
Arachidonic acid release and prostaglandin synthesis	Millar & Rozengurt (1990)
Enhancement of cAMP accumulation	Millar & Rozengurt (1988)
Increase in c-fos and c-myc nRNA levels	Rozengurt & Sinnett-Smith (1988); Mehmet et al (1989a,b)
Elevation of c-fos protein	
Stimulaton of DNA synthesis	Rozengurt & Sinnett-Smith (1983)

specific synergistically acting signal transduction pathways initiated almost immediately after mitogen addition (Rozengurt, 1986; Rozengurt et al, 1988).

Neuropeptides are increasingly recognised to act as cellular growth factors (Zachary et al, 1987) and their mechanisms of action are attracting considerable attention. In particular, bombesin (Rozengurt and Sinnett-Smith, 1983), vasopressin (Rozengurt et al, 1979), bradykinin (Woll and Rozengurt, 1988), vasoactive intestinal peptide (Zurier et al, 1988), endothelin (Takuwa et al, 1989) and vasoactive intestinal contractor (Fabregat and Rozengurt, 1990) can act as growth factors for cultured 3T3 cells. In what follows, some fundamental features of the mechanism of action of neuropeptides as growth factors in 3T3 cells will be discussed. Subsequently, the evidence for multiple neuropeptide growth factor action in lung tumours will be considered before discussing the development of drugs directed against them.

EARLY SIGNALLING EVENTS

The early cellular and molecular responses elicited by bombesin and structurally related peptides in 3T3 cells (Table 1) have been elucidated in detail (Rozengurt and Sinnett-Smith, 1990). The cause-effect relationships and temporal organization of these early signals and molecular events provide a paradigm for the study of other growth factors and mitogenic neuropeptides and illustrate the activation and interaction of a variety of signalling pathways (Rozengurt, 1991).

Bombesin/GRP binds to a single class of high affinity receptors in Swiss 3T3 cells (Zachary and Rozengurt, 1985; Sinnett-Smith et al, 1990). The receptor is coupled to one or more guanine nucleotide binding proteins (G proteins) as judged by the modulation of ligand binding in either membrane preparations or in receptor solubilized preparations and of signal transduction in permeabilized cells (Erusalimsky et al, 1988; Coffer et al, 1990; Sinnett-Smith et al, 1990; Rozengurt et al, 1990). The bombesin/GRP receptor has recently been cloned and sequenced (Battey et al, 1990; Spindel et al, 1990) and shown to be a member of the G protein coupled receptor family. These receptors have seven predicted transmembrane domains which cluster to form a ligand-binding pocket (Dohlman et al, 1987; Lefkowitz and Caron, 1988). Other neuropeptide mitogens with receptors of this type include angiotensin, endothelin, serotonin, substance K and substance P (Arai et al, 1990; Sakurai et al, 1990).

Binding of bombesin/GRP to its receptor initiates a cascade of intracellular signals (summarized in Table 1) culminating in DNA synthesis 10-15 h later (Rozengurt 1986; 1991). One of the earliest events to occur after the binding of bombesin to its specific receptor is a rapid mobilization of Ca^{2+} from internal stores, which leads to a transient increase in the intracellular concentration of Ca^{2+} ($[Ca^{2+}]_i$) and subsequently to Ca^{2+} efflux and decreased Ca^{2+} content of the cells (Mendoza et al, 1986; Lopez-Rivas et al, 1987). The mobilization of Ca^{2+} by bombesin is mediated by inositol 1,4,5-trisphosphate $[Ins(1,4,5)P_3]$, which, as a second messenger binds to an intracellular receptor and induces the release of Ca^{2+} from internal stores. Bombesin causes a rapid increase in $Ins(1,4,5)P_3$, which coincides with the increase in cytosolic Ca^{2+} (Nanberg and Rozengurt, 1988). $Ins(1,4,5)P_3$ is formed as a result of phospholipase C (PLC) catalysed hydrolysis of phosphatidyl inositol 4,5-bisphosphate (PIP_2) in the plasma membrane, a process that also generates 1,2-diacyl-glycerol (DAG). DAG can also be generated from other sources, such as phosphatidylcholine hydrolysis (Cook, and Wakelam, 1989), and acts as a second messenger in the activation of protein kinase C (PKC) by bombesin. In accord with this, bombesin strikingly increases the phosphorylation of the acidic 80 K protein (Rozengurt et al, 1983; Zachary et al, 1986; Erusalimsky et al, 1988), a major substrate of PKC which has been recently purified from Swiss 3T3 cells (Brooks et al, 1990) and molecularly cloned (Brooks et al, 1991; Erusalimsky et al, 1991). Bombesin/GRP also stimulates a rapid exchange of Na^+, H^+ and K^+ ions across the cell membrane, leading to cytoplasmic alkalinization and increased intracellular $[K^+]$ (Mendoza et al, 1986) and induces a striking PKC-dependent transmodulation of the epidermal growth factor receptor (Zachery et al, 1986).

Recently, bombesin, vasopressin and endothelin have been shown to induce a rapid and potent stimulation of tyrosine phosphorylation of several substrates in quiescent 3T3 cells (Zachery et al, 1991). This response is not mediated by either PKC activation or Ca^{2+} mobilization. The mechanism by which neuropeptide receptors elicit this novel pathway as well as the precise role of tyrosine phosphorylation in neuropeptide mediated signal transduction are intriguing issues that warrant further experimental work.

In addition, bombesin, but not vasopressin, induces a marked and sustained release of arachidonic acid and its cyclo-oxygenase metabolite PGE_2 into the medium (Millar and Rozengurt, 1990). Considerable evidence indicates that the liberation of arachidonic acid is an early signal that contributes to bombesin-mediated mitogenesis (Millar and Rozengurt, 1990; Gil et al, 1991).

In common with many other growth factors, bombesin/GRP stimulates
transient expression of the nuclear oncogenes c-fos and c-myc (Rozengurt
and Sinnett-Smith, 1988). It is likely that the induction of c-fos by
bombesin is mediated by the co-ordinated effects of PKC activation, Ca^{2+}
mobilization and an additional pathway dependent on arachidonic acid
release (Rozengurt and Sinnett-Smith, 1988; Mehmet et al, 1990a,b).
Furthermore, additional pathways of control of c-fos expression that are
completely independent of activation of PKC have also been shown.
Indeed, bombesin can initiate DNA synthesis via PKC-dependent and -
independent pathways (Rozengurt and Sinnett-Smith, 1988). This complex
network of signals involves a degree of redundancy, and ensures the
amplification of the stimulus.

In addition to bombesin, several other regulatory peptides have been
characterized as mitogens for Swiss 3T3 cells including vasopressin,
bradykinin and endothelin-related peptides, and their signalling pathways
have also been defined in detail. These neuropeptide receptors are also
linked to phosphoinositide breakdown and Ca^{2+} mobilization but the
intensity, duration (e.g. PKC activation) and even the occurrence of
early signals (e.g. arachidonic acid release) differ substantially
(Millar and Rozengurt, 1990; Issandou and Rozengurt, 1990; Rozengurt,
1991).

MULTIPLE NEUROPEPTIDES STIMULATE Ca^{2+} MOBILIZATION AND CLONAL GROWTH IN
SCLC CELLS

Lung cancer is the commonest fatal malignancy in the developed
world. SCLC constitutes 25% of the total and follows an aggressive
clinical course, despite initial chemosensitivity (Smyth et al. 1986).
Identification of the factors that stimulate the proliferation of SCLC
cells will be important in the design of alternative and more effective
therapeutic strategies. SCLC is characterized by the presence of
intracytoplasmic neurosecretory granules and by its ability to secrete
many hormones and neuropeptides (Sorenson et al, 1981; Maurer, 1985)
including bombesin, neurotensin, cholecystokinin and vasopressin (North
et al, 1980; Sorenson et al, 1981; Wood et al, 1981; Gazdar et al, 1984;
Goadert et al, 1984; Maurer, 1985; Sausville et al, 1985). Among these,
only bombesin-like peptides, which include gastrin-releasing peptide
(GRP), have been shown to act as autocrine growth factors for certain
SCLC cell lines (Cuttitta et al, 1985; Mahmoud et al, 1991).
Consequently, it is important to elucidate the role played by other
neuropeptides in SCLC growth.

Ca^{2+} mobilization is one of the components of a complex array of
signalling events leading to cell growth in Swiss 3T3 cells. GRP also
stimulates mobilization of intracellular Ca^{2+} and inositol phosphate
turnover in SCLC cells (Heikkila et al, 1987; Trepel et al, 1988). In a
subsequent study, Woll and Rozengurt (1989) screened multiple
neuropeptides for their ability to induce a rapid increase in $[Ca^{2+}]_i$ in
different SCLC cell lines. These studies demonstrated that bradykinin,
cholecystokinin, galanin, neurotensin and vasopressin induce a rapid and
transient increase in $[Ca^{2+}]_i$ in a dose-dependent fashion in the
nanomolar range (Sethi and Rozengurt, 1991a,b). The Ca^{2+}-mobilizing
effects are mediated by distinct receptors as shown by the use of
specific antagonists and by the induction of homologous desensitization
(Woll and Rozengurt, 1989; Sethi and Rozengurt, 1991a,b). Studies
carried out in other laboratories are in agreement with these findings
(Staley et al, 1989a,b; Bunn et al, 1990).

In view of the preceding findings it was important to determine whether Ca^{2+}-mobilizing neuropeptides can act as growth factors for SCLC cell lines. Consequently, we determined the effect of multiple Ca^{2+}-mobilizing neuropeptides to promote clonal growth in semi-solid medium in different SCLC cell lines. Sethi and Rozengurt (1991a,b) demonstrated that, at optimal concentrations, bradykinin, neurotensin, vasopressin, cholecystokinin, galanin, and GRP induce comparable increases of SCLC clonal growth in responsive cell lines. Thus, multiple Ca^{2+}-mobilizing neuropeptides, via distinct receptors, can act directly as growth factors for SCLC.

It is known that GRP, vasopressin, cholecystokinin and neurotensin are secreted by some SCLC tumours (Sethi and Rozengurt, 1991a,b). Other peptides may be released by a variety of normal cells in the lung or, like bradykinin, produced extracellularly as a result of the proteolytic cleavage of plasma precursors in the damaged tissue surrounding tumours (Steranka et al, 1989). Collectively, these findings support the hypothesis that SCLC growth is sustained by an extensive network of autocrine and paracrine interactions involving multiple neuropeptides. Approaches designed to block SCLC growth must take into account this mitogenic complexity.

BLOCKING THE ACTION OF MULTIPLE NEUROPEPTIDES: BROAD SPECTRUM ANTAGONISTS

As understanding of the effects of growth factors on cancer increases, it has become possible to plan rational therapeutic interventions. If an autocrine growth loop is considered, in which cells synthesize, secrete, bind and respond to the same growth factor, it is evident that interruption of this cycle at any point will block mitogenesis. Paracrine growth could be blocked in the same way. As discussed in the preceding sections, SCLC constitutes a special case in which unrestrained proliferation appears driven, at least in part, by multiple autocrine and paracrine circuits involving Ca^{2+}-mobilizing neuropeptides.

Secreted factors can be cleared by antibodies, such as the bombesin monoclonal antibody 2A11 used to retard the growth of SCLC xenografts in nude mice (Cuttitta et al, 1985). We have directed our effort to develop peptide antagonists which are not antigenic and should have higher tissue penetration than antibody proteins. We have characterized neuropeptide antagonist in the model Swiss 3T3 fibroblast system and then tested their effects on SCLC in vitro and in vivo.

The first antagonist to be studied was an analogue of substance P, [DArg[1], DPro[2], DTrp[7,9], Leu[11]]substance P (antagonist A, Table 2). Substance P is structurally unrelated to the bombesin-like peptides, but antagonist A, which is a substance P antagonist, was found to block the secretory effects of bombesin on a pancreatic preparation (Jensen et al, 1984). It was subsequently found to block ^{125}I-GRP binding and bombesin-stimulated early signalling events and mitogenesis in Swiss 3T3 cells (Zachary and Rozengurt, 1985; Mendoza et al, 1986; Erusalimsky et al, 1988; Rozengurt and Sinnett-Smith, 1988; Woll and Rozengurt, 1988). It did not affect mitogenesis stimulated by polypeptide growth factors, such as EGF and platelet-derived growth factor (PDGF), but was found to block vasopressin-stimulated mitogenesis (Zachary and Rozengurt, 1986). Further substance P analogues were therefore studied in order to identify more potent antagonists that could be tested in SCLC (Woll and Rozengurt, 1988b; 1990).

Table 2. Bombesin/GRP and broad spectrum antagonists

Bombesin:

pGlu-Gln-Arg-Leu-Gly-Asn-Gln-Trp-Ala-Val-Gly-His-Leu-Met-NH$_2$

Broad spectrum antagonists (substance P analogues)

Susbstance P: Arg-Pro-Lys-Pro-Gln-Gln-Phe-Phe-Gly-Leu-Met-NH$_2$

Antagonist A: DArg-DPro-Lys-Pro-Gln-Gln-DTrp-Phe-Dtrp-Leu-Leu-NH$_2$

Antagonist D: DArg-Pro-Lys-Pro-DPhe-Gln-DTrp-=Phe-DTrp-Leu-Leu-NH$_2$

Antagonist G: Arg-DTrp-MePhe-DTrp-Leu-Met-NH$_2$

Two interesting compounds were [DArg^1, DPhe^5, DTrp^{7,9}, Leu^{11}], substance P (antagonist D, Table 2) and [Arg^6, DTrp^{7,9}, MePhe^8] substance P(6-11) (antagonist G, Table 2). Both antagonists reversibly inhibited GRP-stimulated mitogenesis in Swiss 3T3 cells, and antagonist D was 5-fold more potent than antagonist A, although antagonist G was less potent than A (Woll and Rozengurt, 1990). In contrast, when tested as competitive inhibitors of vasopressin-stimulated mitogenesis, antagonists D and G were equipotent. In addition, the antagonists were found to block mitogenesis stimulated by the neuropeptides bradykinin and endothelin (Woll and Rozengurt, 1988a,b; Fabregat and Rozengurt, 1990b). It is important to note that the antagonists neither block DNA synthesis by PDGF which stimulates Ca^{2+} mobilization through a different mechanism from neuropeptides (i.e. mediated by tyrosine phosphorylation rather than by a G protein) nor inhibit mitogenesis stimulated by vasoactive intestinal peptide which induces cAMP accumulation without Ca^{2+} mobilization. Thus, the substance P analogue antagonists showed broad spectrum specificity against the neuropeptide mitogens bombesin/GRP, vasopressin, bradykinin and endothelin, which act through distinct receptors in Swiss 3T3 cells, but activate common signal transduction pathways (Rozengurt, 1991).

The molecular mechanism by which broad spectrum antagonists interfere with the action of Ca^{2+}-mobilizing neuropeptides remains to be defined. Antagonists D and G competed with the radio-labelled ligands ^{125}GRP, [^{3}H] vasopressin and ^{125}I-endothelin for binding in a dose-dependent fashion and inhibited Ca^{2+} mobilization stimulated by each of these peptides, in addition to other early intracellular signals triggered by them (Sinnett-Smith et al, 1990; Woll and Rozengurt, 1988b, 1990; Fabregat and Rozengurt, 1990b). It is plausible that the antagonists recognize a common domain on these Ca^{2+}-mobilizing neuropeptide receptors, each of which is probably coupled to a common G protein responsible for the regulation of polyphosphoinositide-specific phospholipase C. Alternatively, the antagonists might bind to a separate protein that regulates receptor activity.

Table 3. The effect of multiple peptide hormones and neuropeptides on $[Ca^{2+}]_i$ mobilization in SCLC cell lines

Effective	Non Effective
Bradykinin	ACTH
Cholecystokinin	Angiotensin I, II, III
Galanin	Atrial natriuretic peptide
Bombesin/GRP	Calcitonin
Neurotensin	Chorionic Gonadotrophin
Vasopressin	Dynorphin
	-endorphin
	Endothelin
	Epinephrine
	Follicle stimulating hormone
	GRH
	GIP
	Glucagon
	5-hydroxytryptamine
	Leu-enkephalin
	Neuropeptide-Y
	Parathyroid hormone
	Substance K
	Substance P
	TRH

Intracellular Ca^{2+} was measured in SCLC cell lines NCI H69, H510, H345, H209, H128 with the indicator fura-2/AME. Effective peptides resulted in consistant large responses at nanomolar concentrations, the responses in the various cell lines were heterogeneous (see Woll and Rozengurt, 1989; Sethi and Rozengurt, 1991a,b).

BROAD SPECTRUM ANTAGONISTS BLOCK SCLC GROWTH

The compounds characterized as broad spectrum antagonists in Swiss 3T3 cells were tested as inhibitors of neuropeptide mediated signals and growth in SCLC cell lines. Because SCLC is a heterogeneous group of tumours, each compound was tested in several cell lines. The broad spectrum antagonists inhibited Ca^{2+} mobilization stimulated by GRP, vasopressin, bradykinin, cholecystokinin and galanin in diverse cell lines and inhibited the growth of SCLC cell lines, in liquid and semi-solid media (Woll and Rozengurt, 1990; Sethi and Rozengurt, 1991b). Antagonists D and G were equipotent, with half-maximal effect at about 20 μM, whereas antagonist A was 5-fold less potent.

The broad spectrum antagonists (D and G) caused a dramatic decrease of the cloning efficiency of these cells in the absence of any exogenously added peptide (i.e. basal colony formation). Broad spectrum antagonists also decrease clonal growth in the presence of neuropeptide stimulation (Sethi and Rozengurt, 1991b). For example, antagonist G profoundly inhibited the clonal growth of SCLC H69 or H345 cells in the absence, as well as in the presence of either galanin or vasopressin (Sethi and Rozengurt, unpublished results). The striking finding that antagonists D and G inhibit the basal and stimulated clonal growth of so many cell lines (Woll and Rozengurt, 1990; Sethi and Rozengurt, 1991b), regardless of positivity for bombesin receptors, suggests that broad spectrum antagonists could be more useful anticancer drugs than ligand specific growth factor antagonists.

CONCLUSIONS

Neuropeptides are increasingly implicated in the control of cell proliferation and their mechanisms of action are attracting intense interest. The peptides of the bombesin family including gastrin-releasing peptide (GRP) bind to specific surface receptors and initiate a complex cascade of signalling events (Table 1) that culminates in the stimulation of DNA synthesis and cell division in Swiss 3T3 cells in the absence of other growth promoting factors. These peptides may also act as autocrine growth factors for certain SCLC cells. The results discussed here strongly suggest that the autocrine growth loop of bombesin-like peptides may be only a part of an extensive network of autocrine and paracrine interactions involving a variety of Ca^{2+}-mobilizing neuropeptides in SCLC including bradykinin, cholecystokinin, galanin, neurotensin and vasopressin (Table 3). In the context of the multistage evolution of cancer, neuropeptide mitogenesis may play a role at an early stage in SCLC as tumour promoters in initiated cells or later as growth factors in the unrestrained growth of the fully developed SCLC tumour. A detailed understanding of the receptors and signal transduction pathways that mediate the mitogenic action of neuropeptides may identify novel targets for therapeutic intervention. In this context, broad spectrum antagonists that prevent the function of multiple Ca^{2+}-mobilizing receptors are of special interest. These antagonists block neuropeptide mediated signals in the 3T3 and SCLC cells and inhibit SCLC growth. Thus, broad spectrum neuropeptide antagonists constitute potential anticancer agents.

REFERENCES

Arai, H., Hori, S., Aramori, I., Ohkubo, H. and Nakanishi, S., 1990,
    Cloning and expression of a cDNA encoding and endothelin
    receptor, Nature, 348:730.
Battey, J.F., Way, J.M., Corjay, M.H., Shapira, H., Kusano, K.,
    Harkins, R., Wu, J.M., Slattery, T., Mann, E. and Feldman, R.I.,
    1990, Molecular cloning of the bombesin/GRP receptor from Swiss
    3T3 cells, Proc. Natl. Acad. Sci. USA, 88:395.
Bishop, J.M., 1991, Molecular themes in oncogenesis, Cell,64:235.
Brooks, S.F., Erusalimsky, J.D., Totty, N.F. and Rozengurt, E., 1990,
    Purification and internal amino acid sequence of the 80 kDa
    protein kinase C substrate from Swiss 3T3 fibroblasts,
    FEBS Letters, 268:291.
Brooks, S.F., Herget, T., Erusalimsky, J.D. and Rozengurt,E., 1991,
    Protein kinase C activation potently down-regulates the expression
    of its major substrate, 80K, in Swiss 3T3 cells, EMBO J.,
    (in press).
Bunn, P.A., Dienhart, D.G., Chan, D., Puck, T.T., Tagawa, M., Jewett,
    P.B. and Braunschweiger, E., 1990, Neruopeptide stimulation of
    calcium flux in human lung cancer cells: delineation of
    alternative pathways, Proc. Natl. Acad. Sci. USA, 87:2162.
Coffer, A., Fabregat, I., Sinnett-Smith, J. and Rozengurt, E., 1990,
    Solubilization of the bombesin receptor from Swiss 3T3 cells
    membranes: functional association to a guanine nucleotide
    regulatory protein, FEBS Lett., 263:80.
Cook, S.J. and Wakelam, J.O., 1989, Analysis of the water-soluble

products of phosphatidylcholine breakdown by ion-exchange chromatography, Biochem. J., 263:581.

Cross, M. and Dexter, T.M., 1991, Growth factors in development, transformation, and tumorigenesis, Cell, 64:271.

Cuttitta, F., Carney, D.N., Mulshine, J., Moody, T.W., Fedorko, J., Fischler, A. and Minna, J.D., 1985, Bombesin-like peptides can function as autocrine growth factors in human smal-cell lung cancer, Nature, 316:823.

Dohlman, H.G., Caron, M.G.. and Lefkowitz, R.J., 1987, A family of receptors coupled to guanine nucleotide regulatory proteins, Biochem. J., 26:2657.

Erusalimsky, J.D., Friedberg. I. and Rozengurt, E., 1988, Bombesin, diacylglycerols and phorbol esters rapidly stimulate the phosphorylation of an Mr = 80,000 protein kinase C substrate in permeabilized 3T3 cells: effect of guanine nucleotides, J. Biol. Chem., 263:19188.

Erusalimsky, J.D., Brooks, S.F., Herget, T., Morris, C. and Rozengurt, E., 1991, Molecular cloning and characterization of the acidic 80 kDa protein kinase C substrate from rat brain, J. Biol. Chem., 266:7073.

Fabregat, I. and Rozengurt, E., 1990a, Vasoactive intestinal contractor, a novel peptide, shares a common receptor with endothelin-1 and stimulates Ca^{2+} mobilization and DNA synthesis in Swiss 3T3 cells, Biochem. Biophys. Res. Commun., 167:161.

Fabregat, I. and Rozengurt, E., 1990b, [$DArg^1$,$DPhe^5$,$DTrp^{7,9}$,Leu^{11}] substance P, a neuropeptide antagonist, blocks binding, Ca^{2+}-mobilizing, and mitogenic effects of endothelin and vasoactive intestinal contractor in mouse 3T3 cells, J. Cell. Physiol., 145:88.

Gazdar, A.F. and Carney, D.N., 1984, in: "The endocrine Lung in Health and Disease," K. Becker and A.F. Gazdar, eds., W.B. Saunders, London.

Gil, J., Higgins, T. and Rozengurt, E., 1991, Mastoparan, a novel mitogen for swiss 3T3 cells, stimulates pertussis toxin-sensitive arachidonic acid release without inosital phosphate accumulation, J. Cell Biol., 113, 943.

Goedert, M., Reeve, J.G., Emson, P.C. and Bleehen, N.M., 1984, Neuro-tensin in human small cell lung carcinoma, Br. J. Cancer, 50:179.

Goustin, A.S., Leof, E.B., Shipley, G.D. and Moses, H.L., 1986, Growth factors and cancer, Cancer Res., 46:1015.

Heikkila, R., Trepel, J.B., Cuttitta, F., Newckers, L.M. and Sausville, E.A., 1987, Bombesin-related peptides induce calcium mobilization in a subset of human small cell lung cancer cell lines, J. Biol. Chem., 262:16456.

Issandou, M. and Rozengurt, E., 1990, Bradykinin transiently activates protein kinase C in Swiss 3T3 cells: distinction from activation by bombesin and vasopressin, J. Biol. Chem., 265:11890.

Jensen, R.T., Jones, S.W., Folkers, K. and Gardner, J.D., 1984, A synthetic peptide that is a bombesin receptor antagonist, Nature, 309:61.

Kris, R.M., Hazan, R., Villines, J., Moody, T.W. and Schlessinger, J., 1987, Identification of the bombesin receptor on murine and human cells by cross-linking experiments, J. Biol. Chem., 262:11215.

Lefkowitz, R.J. and Caron., M.G., 1988, Adrenertic receptors: models for the study of receptors coupled to guanine nucleotide regulatory proteins, J. Biol. Chem., 263:4993.

Lopez-Rivas, A., Mendoza, S.A., Nanberg, E., Sinnett-Smith, J. and Rozengurt, E., 1987, The Ca^{2+}-mobilizing actions of platelet-derived growth factor differ from those of bombesin and vasopressin in Swiss 3T3 cells, Proc. Natl. Acad. Sci. USA, 84:5768.

Mahmoud, S., Staley, J., Taylor, J., Bogden, A., Moreau, J.-P., Coy, D., Avis, I., Cuttitta, F., Mulshine, J.L. and Moody, T.W., 1991, [Psi[13,14]] bombesin analogues inhibit growth of small cell lung cancer in vitro and in vivo, Cancer Res., 51:1789.

Maurer, L.H., 1985, Ectopic hormone syndrome in small cell carcinoma of the lung, Clinics. Oncol., 4:1289.

Mehmet, H., Morris, C. and Rozengurt, E., 1990a, Multiple synergistic signal transduction pathways regulate c-fos expression in swiss 3T3 cells: the role of cyclic AMP, Cell. Growth. Diff., 1:292.

Mehmet, H., Millar, J. B. A., Lehmann, W., Higgins., T. and Rozengurt, E., 1990b, Bombesin stimulation of c-fos expression and mitogenesis in Swiss 3T3 cells: The role of prostaglandin E_2-meditated cyclic AMP accumulation, Exp. Cell Res., 190:265.

Mendoza, S. A., Schneider, J. A., Lopez-Rivas, A., Sinnett-smith, J. W. and Rozengurt, E., 1986, Early events elicited by bombesin and structurally related peptides in quiescent Swiss 3T3 cells, II. Changes in Na^+ and Ca^{2+} fluxes, Na^+/K^+ pump activity and intracellular pH, J. Cell Biol., 102:2223.

Millar, J.B.A. and Rozengurt, E., 1990, Arachidonic acid release by bombesin: a novel post-receptor target for heterologous mitogenic desensitization, J. Biol. Chem., 265:19973.

Nanberg, E. and Rozengurt, E., 1988, Temporal relationship between inositol polyphosphate formation and increases in cytosolic Ca^{2+} in quiescent 3T3 cells stimulated by platelet-derived growth factor, bombesin and vasopressin, EMBO. J., 7:2741.

North, W.G., Maurer, L.H., Valtin, H. and O'Donnell, J.F., 1980, Human neurophysins as potential tumor markers for small cell carcinoma of the lung: application of specific radioimmunoassays, J. Clin. Endocrinol. Metab., 51:892.

Rozengurt, E., 1983, Growth factors, cell proliferation and cancer: An overview, Mol. Biol. Med., 1:169.

Rozengurt, E., 1985, The mitogenic response of cultured 3T3 cells: integration of early signals and synergistic effects in a unified framework, in: "Molecular Mechanisms of transmembrane signalling," Cohen P, Houselay M. eds., Amsterdam: Elsevier Science Publishers B.V. Rozengurt, E., 1986, Early signals in the mitogenic response, Science, 234:161.

Rozengurt, E., 1991, Neuropeptides as cellular growth, Eur. J. Clin. Invest, 21:123.

Rozengurt, E. and Sinnett-Smith, J., 1983, Bombesin stimulaton of DNA synthesis and cell division in cultures of Swiss 3T3 cells, Proc. Natl. Acad. Sci, USA, 80:2936.

Rozengurt, E. and Sinnett-Smith, J., 1988, Early signals underlying the induction of the c-fos and c-myc genes in quiescent fibroblasts: studies with bombesin and other growth factors. Prog. Nucl. Acid. Res. Mol. Biol., 35:261.

Rozengurt, E. and Sinnett-Smith, J., 1990, Bombesin stimulation of fibroblast mitogenesis: specific receptors, signal transduction and early events, Phil. Trans. Roy. Soc. London B,, 327:209.

Rozengurt, E., Legg, A. and Pettican P., 1979, Vasopressin stimulation of 3T3 cell growth, Proc. Natl. Acad. Sci, USA, 76:1284.

Rozengurt, E., Rodriguez-Pena, A. and Smith, K. A., 1983, Phorbolesters, phopholipase C, and growth factors rapidly stimulate the phosphorylation of a M_r 80,000 protein in intact quiescent 3T3 cells, Proc. Natl. Acad. Sci, USA, 80:7244.

Rozengurt, E., Erusalimsky, J., Mehmet, H., Morris, C., Nanberg, E. and Sinnett-Smith, J., 1988, Signal transduction in mitogenesis: further evidence for multiple pathways, Cold Spring Harbor Symp. Quant. Biol., 53:945.

Rozengurt, E., Fabregat, I., Coffer, A., Gil, J. and Sinnett-Smith, J., 1990, Mitogenic signalling through the bombesin receptor: role of a guanine nucleotide regulatory protein, J. Cell. Sci. Suppl. 13:43.

Sager, R., 1989, Tumor suppressor genes: The puzzle and the promise, Science, 246:406.

Sakurai, T., Yanagisawa, M., Takuwa, Y., Miyazaki, H., Kimura, S., Goto, K. and Masaki, T., 1990, Cloning of a cDNA encoding a non-isopeptide-selective subtype of the endothelin receptor, Nature, 348:732.

Sausville, E., Carney, D and Battey, J., 1985, The human vasopressin gene is linked to the oxytocin gene and is selectively expressed in a cultured lung cancer cell line, J. Biol. Chem., 260:10236.

Sethi, T. and Rozengurt, 1991, Multiple neuropeptides stimulate clonal growth of small cell lung cancer: Effects of bradykinin, vasopressin, cholecystokinin, galanin and neurotensin, Cancer Res., 51:3621.

Sinnett-Smith, J., Lehmann, W. and Rozengurt, E., 1990, Bombesin receptor in membranes from Swiss 3T3 cells. Binding characteristics, affinity labelling and modulation by guanine nucleotides, Biochem. J, 265:485.

Sinnett-Smith, J., Zachary, I. and Rozengurt, E., 1988, Characterization of a bombesin receptor on Swiss mouse 3T3 cells by affinity cross-linking, J. Cell. Biochem., 38:237.

Smyth, J.F., Fowlie, S.M., Gregor, A., Crompton, G.K., Busutill, A., Leonard, R.C.F. and Grant, I.W.B., 1986, The impact of chemotherapy on small cell carcinoma of the bronchus, Quart. J. Med., 61:969.

Sorenson, G.D., Pettengill, O.S., Brinck-Johnsen, T., Cate, C.C. and Maurer, L.H., 1981, Hormone production by cultures of small-cell carcinoma of the lung, Cancer, 47:1289.

Spindel, E.R., Giladi, E., Brehm, P., Goodman, R.H. and Segerson, T.P., 1990, Cloning and functional characterization of a complementary DNA encoding the muring fibroblast bombesin/gastrin-releasing peptide receptor, Mol. Endocrinol., 4:1956.

Sporn, M.B. and Roberts, A.B., 1985, Autocrine growth factors and cancer, Nature, 313:745.

Staley, J., Fiskum., G. and Moody, T.W., 1989, Cholecystokinin elevates cytosolic calcium in small cell lung cancer cells, Biochem. Biophys. Res. Commun., 1643:605.

Steranka, L.R., Farmer, S.G. and Burch, R.M., 1989, Antagonists of bradykinin receptors, FASEB J, 3:2019.

Takuwa, N., Takuwa, Y., Yanagisawa, M., Yamashita, K. and Masaki, T., 1989, A novel vasoactive peptide endothelin stimulates mitogenesis through inositol lipid turnover in Swiss 3T3 fibroblasts, J. Biol. Chem, 264:7856.

Trepel, J.B., Moyer, J.D., Heikkila, R. and Sausville, E.A., 1988, Modulation of bombesin-induced phosphatidylinositol hydrolysis in a small-cell lung-cancer cell line, Biochem. J, 255:403.

Woll, P. J. and Rozengurt E., 1988a, Two classes of antagonist interact with receptors for the mitogenic neuropeptides bombesin, bradykinin and vasopressin, Growth Factors, 1:75.

Woll, P.J. and Rozengurt, E., 1988b, [DArg^1, DPhe^5, DTrp^{7,9}, Leu^{11}] substance P, a potent bombesin antagonist in murine Swiss 3T3 cells, inhibits the growth of human small cell lung cancer cells in vitro, Proc. Natl. Acad. Sci. USA, 85:1859.

Woll, P.J. and Rozengurt, E., 1989, Multiple neuropeptides mobilise calcium in small cell lung cancer: effects of vasopressin, bradykinin, cholecystokinin, galanin and neurotensin, Biochem. Biophys. Res. Commun, 164:66.

Woll, P.J. and Rozengurt, E., 1990, A neuropeptide antagonist that inhibits the growth of small cell lung cancer in vitro, Cancer Res, 50:3968.

Wood, S.M., Wood, J.R., Ghatei, M.A., Lee, Y.C., O'Shaughnessy, D. and Bloom, S.R., 1981, Bombesin, somatostatin and neurotensin-like immunoreactivity in bronchial carcinoma, J. Clin. Endocrinol. Metab, 53:1310.

Zachary, I. and Rozengurt, E., 1985, High-affinity receptors for peptides of the bombesin family in Swiss 3T3 cells, Proc. Natl. Acad. Sci, USA, 82:7616.

Zachary, I. and Rozengurt, E., 1986, A substance P antagonist also inhibits the specific binding and mitogenic effects of vasopressin and bombesin-related peptides in Swiss 3T3 cells, Biochem. Biophys. Res. Commun, 137:135.

Zachary, I. and Rozengurt, E., 1987, Identification of a receptor for peptides of the bombesin family in Swiss 3T3 cells by affinity cross-linking, J. Biol. Chem, 262:3947.

Zachary, I., Sinnett-Smith, J.W. and Rozengurt, E., 1986, Early events elicited by bombesin and structurally related peptides in quiescent Swiss 3T3 cells. I. Activation of protein kinase C and inhibition of epidermal growth factor binding, J. Cell Biol, 124:295.

Zachary, I., Gil, J., Lehmann, W., Sinnett-Smith, J. and Rozengurt, E.,1991, Bombesin, vasopressin and endothelin rapidly stimulate tyrosine phosphorylation in intact Swiss 3T3 cells, Proc. Natl. Acad. Sci. USA, 88: 4577.

Zurier, R. B., Kozma M., Sinnett-Smith, J. and Rozengurt, E., 1988, Vasoactive intestinal peptide synergistically stimulates DNA synthesis in mouse 3T3 cells: role of cAMP, Ca^{2+} and protein kinase C, Exp. Cell Res., 176:155.

BACTERIAL VECTORS TO TARGET AND/OR PURIFY POLYPEPTIDES

M. Hofnung, A. Charbit, J.-M. Clément, C. Leclerc[*],
P. Martineau, S. Muir, D. O'Callaghan, O. Popescu,
and S. Szmelcman

Unité de Programmation Moléculaire and
Toxicologie Génétique (CNRS URA 1444) and
*Biologie des Régulations Immunitaires, Institut Pasteur
25 rue du Dr Roux, 75015 Paris, France

INTRODUCTION

The construction of recombinant proteins by genetic engineering has opened new avenues in basic research (studies on protein organization, protein folding, immunogenicity of proteins, etc) and in the development of applications (vaccines, diagnosis, targeting of polypeptides, combination of activities, etc). Recombinant proteins which retain properties of both parental proteins are especially interesting. For example, if one protein (the vector protein) is targeted to a given cellular compartment, the other protein (the passenger) may be identically targeted. Also, if the vector protein can be purified by a simple affinity chromatographic procedure, this property may be extended to the passenger. However, recombinant proteins constructed without precaution are often unstable, toxic, insoluble or have lost some of the useful properties of the "parent" proteins.

We have developed a genetic procedure to detect "permissive" sites within potential vector proteins so that genetic insertions into, or fusions to, these sites preserve most or all biological properties of the vector. A number of perspectives are opened by this approach. (1) Study of membrane protein topology by genetic insertion of a foreign epitope and probing with a single monoclonal antibody. (2) Study of the immune response to a passenger peptide depending on its site of insertion in a vector protein and on the bacterial cellular compartment to which it is targeted, with implications for recombinant vaccines. (3) One step purification of functional domains from procaryote or eucaryote proteins expressed in E. coli, including the soluble part of the CD4 receptor for the human immunodeficiency virus (HIV). (4) Induction and monitoring of anti-peptide immunodeficiency antibodies without peptide synthesis.

Proteins are often able to perform complex tasks. For example, the primary product of the gene iga from the bacterium Neisseria gonorrhoeae is a precursor protein of about 169 kd which contains all the instructions to be selectively secreted not only by Neisseria but also by other bacterial species such as E. coli (Klauser et al., 1990 and references therein) and to yield the mature IgA protease (106kd) able to cleave

Targeting of Drugs 3: The Challenge of Peptides and Proteins
Edited by G. Gregoriadis et al., Plenum Press, New York, 1992

109

specifically IgA antibodies. The precursor consists of four distinct domains: (i) an amino-terminal signal peptide involved in the export through the inner membrane; (ii) the protease domain; (iii) the so-called alpha domain, an extremely basic alpha-helical region which is secreted with the protease; (iv) the beta domain which carries the essential functions for transport through the outer-membrane of these Gram- bacteria.

Thus a single natural polypeptide chain may carry all the instructions to perform its catalytic functions at the right cellular location. This means that it carries also the instructions for the targeting of the catalytic activity in the celullar context. These combinations of instructions, often corresponding to combinations of domains, are supposed to result from a long and selective evolutionary process.

We have been interested in the following question. Is it possible to add a domain to a protein by genetic means and obtain a recombinant protein with some or all properties of both parents? This question is not a trivial one because very often recombinant proteins expressed in bacteria create problems. For example they can be inactive, toxic, unstable, insoluble, etc.

In this chapter, we discuss a general genetic method to find and use sites in a "vector" protein where foreign "passenger" domains can be inserted so that the recombinant protein retains most or all properties of the parents. We used mainly two bacterial envelope proteins as vectors. These two proteins are exported beyond the cytoplasmic membrane: LamB is an outer membrane protein from E. coli, with which foreign sequences can be expressed at both sides of the outer membrane; MalE is a periplasmic maltose binding protein from E. coli, with which passengers can be exported to the periplasm and purified in one-step affinity chromatography in mild non-denaturing conditions.

LamB AND EXPRESSION AT THE BACTERIAL SURFACE

The LamB protein is located in the outer membrane of E. coli K12. The active form is a trimer. LamB acts as a general pore which allows the non-specific diffusion of low M.W. hydrophilic molecules (<700). In addition, LamB displays two main specific biological functions: it serves as a receptor for a number of bacteriophages including lambda, and is involved in the penetration of maltose and maltodextrins (Charbit et al., 1988a and therein).

We have developed a general genetic procedure to express foreign epitopes as inserts within "permissive" sites of recipient proteins (Hofnung et al., 1988). These sites are called permissive because they accept genetic modifications without major deleterious consequences for the stability, bacterial toxicity, activities and cellular location of the protein (Charbit et al., 1986). Permissive sites can be identified by a two step procedure. In the first step, mutant proteins generated by in vitro linker mutagenesis are simply screened for conservation of activity (phage receptor and/or dextrins utilization in the case of LamB). In the second step, an oligonucleotide encoding the foreign epitope is inserted in frame into these sites.

Eleven permissive sites have been identified in the LamB protein (Charbit et al., 1991). Insertion at three sites lead to the cell surface exposure of the foreign epitope (residues 153, 253, 374), while

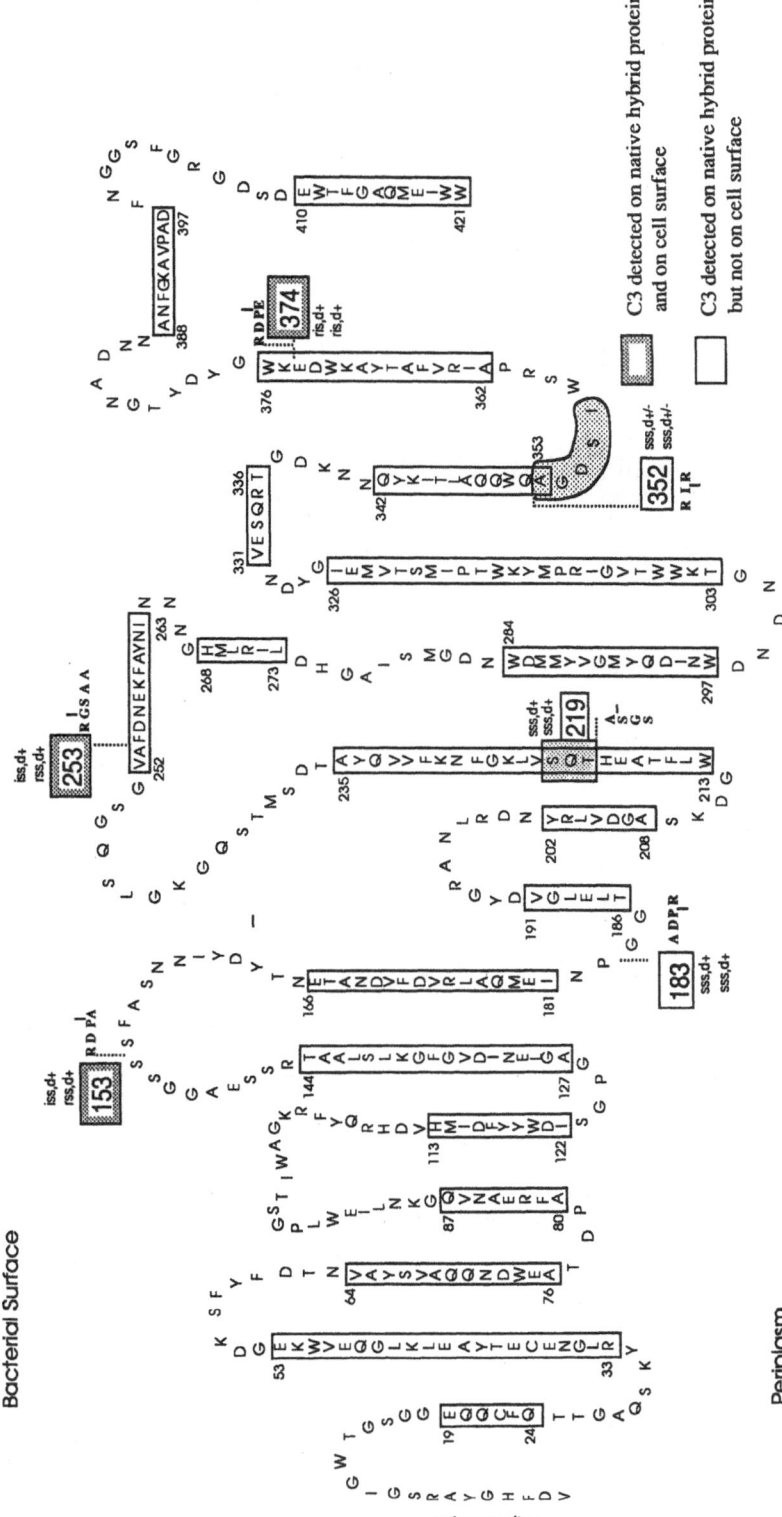

Fig. 1. Six permissive sites in the LamB protein. The sequence of the mature protein is represented in the one letter amino-acid code. On this folding model the LamB protein spans the membrane 16 times. The regions predicted in beta-configuration are presented in rectangles.

at three other sites, the epitope is exposed in the periplasmic side of the membrane (sites 183, 219, 352) (Fig. 1).

We showed that a broad variety of foreign sequences in terms of length and sequence could be expressed at the surface of E. coli by insertion into site 153 of LamB. The hybrid proteins essentially keep their biological activities with inserts up to about 60 residues (Charbit et al., 1988b).

This type of vector may lead to a number of applications.

Live Bacterial Vaccines

The immunogenicity of several viral epitopes, genetically coupled to LamB, was explored (Hofnung, 1988; Leclerc et al., 1989; O'Callaghan et al., 1990; Charbit et al., 1990; Leclerc et al., submitted).

Experiments with recombinant E. coli: Recombinant E. coli expressing the hybrid proteins, as well as purified hybrid proteins were immunogenic and induced high titre antibody responses against the inserted epitopes. The amount of foreign peptide injected could be 100 fold less than that used with synthetic peptide conjugates.

These experiments revealed that several parameters influence the humoral response to the inserted epitope including the route of administration, the preparation of the immunogen, and the location on the carrier protein. By the i.v. route, anti-foreign epitope antibody responses are induced with live, heat killed or sonicated recombinant bacteria. By the s.c. route, anti-foreign epitope antibody responses are induced only when soluble proteins are used as immunogens (semi-purified protein solubilized in detergent).

We were able to study the influence of the exact location of the foreign epitope within LamB and with respect to the outer-membrane. Preliminary results indicate that the position with respect to the membrane plays an important role in determining the intensity and the quality (isotypes) of the antibody response.

Experiments with recombinant Salmonella: Attenuated derivatives of pathogenic invasive bacteria like Salmonella and Shigella have been shown to provide protection upon oral administration in mice against a challenge with the virulent strain (O'Callaghan et al., 1990 and references therein). By using such strains to express heterologous antigens it has become possible to develop multivalent live vaccines.

We showed that the LamB hybrid proteins constructed in E. coli can be expressed at high levels in attenuated strains of Salmonella. First, immunization experiments with recombinant Salmonella indicated that they could be efficient for the induction of anti-foreign epitope antibodies (O'Callaghan et al., 1990). We are currently setting up optimal conditions to achieve a stable and adequate level of expression of recombinant proteins in Salmonella after their delivery into recipient mice.

Simple Diagnostic Tests

We showed that recombinant E. coli expressing foreign peptides at their surface can provide a convenient reagent to monitor and characterize specific antibodies (Charbit et al., 1990). Different peptides from the envelope of HIV1 were expressed at site 153, and the recombinant clones were used to study the immunogenicity of the foreign peptides as well as to test sera from seropositive individuals in Western-blot. The spectrum of LamB hybrid proteins detected was a characteristic of each serum indicating that this technique could possibly be useful for refined diagnosis or even for prognosis. In one case, a serum that was negative by classical tests (ELAVIA and Western-blot) reacted with at least three LamB hybrids. Thus, recombinant bacteria expressing foreign epitopes on their surface may provide sensitive, easy to prepare inexpensive reagents for diagnosis.

Anti-Peptide Antibody

Genetic insertion of the same foreign epitope in LamB and MalE (see below) proteins can provide a simple and flexible procedure to elicit and assay anti-peptide antibodies without peptide synthesis (Martineau et al., 1991). Indeed, one hybrid can be used as immunogen for the induction of antibodies against the inserted sequence, and the other as antigen for monitoring the anti-peptide antibodies raised. These two recipient proteins possess convenient properties: MalE hybrids can be easily purified and LamB hybrids can express the foreign peptide on the cell surface so that intact bacteria can be used a reagents.

Peptides Libraries

By expressing a chosen peptide on the surface of E. coli, it should be possible to select mutants in the peptide with altered recognition for a specific ligand. This would be a similar situation to that described with filamentous phages (Scott et al., 1990; Cwirla et al., 1990). For example, in the case of an epitope-specific antibody combination, it should be possible to isolate mutants with enhanced affinity for the antibody by simple screening procedures. This approach could possibly be extended for protein domains such as enzymes, antibodies, or receptors. Various screening procedures are applicable to this problem (Charbit, Chafotte, Goldberg, Siccardi, Hofnung, unpublished).

MalE EXPRESSION, EXPORTATION AND PURIFICATION VECTOR

MalE is a periplasmic binding protein involved in the transport of maltose and maltodextrins in Escherichia coli. This protein is synthesized in the cytoplasm and exported into the periplasm. It binds maltose and maltodextrins with high affinity. It can be purified by a simple and efficient procedure which involves binding onto an amylose column and elution in mild conditions with maltose. Recently, its structure has been determined at high resolution (Spurlino et al., 1988; Martineau et al., 1990; Spurlino et al., 1991).

We have determined "permissive" sites in this protein (Hofnung et al, 1988). We have analysed in detail insertions in two such sites on MalE, one after residue 133 and the other after residue 303 of the protein. These sites can accommodate peptide insertions up to 80 amino acids without major deleterious effects on protein expression, localization and function. At site 133 all the hydrophilic insertions retain a Mal[+] phenotype and the ability to bind to a cross-linked amylose column. In addition, at this site, it was possible to insert and express in the periplasmic space a peptide as hydrophobic as the signal sequence of the MalE protein. The versatility of the system allowed us to insert, in the continuity of MalE, a large variety of peptides of immunological interest (Martineau et al., 1991).

MalE for Delivery of Peptides to the Immune System

The immunogenicity of two viral epitopes inserted in two sites of MalE was extensively analysed. The first one is a neutralizing B-cell epitope from poliovirus type 1 and the second a dominant B-cell epitope from the pre-S(2) region of hepatitis B virus. The immunologic activities of the hybrid proteins were analysed not only with purified protein but also with live bacteria expressing the hybrids in the periplasm. These studies showed that it is possible to induce high titers of antibodies not only against the peptide but also against the native viral proteins. In addition, neutralizing titers were obtained with the poliovirus epitope (Leclerc et al., 1990).

One of the major features of the MalE system is that the peptide is inserted within the continuity of the protein. Such insertions, compared to classical coupling of peptide either chemically or by C-terminal fusions, limit the mobility of the inserted peptide. We are currently investigating the effects of this constraint, using either an approach with polyclonal and monoclonal antibodies directed against the peptide or by 3-D modeling of the peptide structure by molecular dynamic and energy minimization techniques, or by solving the structure of hybrids by X ray analysis (Rodseth et al., 1990). It was shown that depending on the insertion site and on the peptide, the antigenicity was greatly affected. For example, the poliovirus epitope C3 was more antigenic with a monoclonal antibody at site 133 than at site 303, possibly because its structure is closer to the structure in the native viral particles (Martineau et al., in preparation).

Such studies should help to correlate the antigenicity and the immunogenicity to the 3-D structure of a peptide. By controlling the structure and context of a peptide, it may be possible to induce a response directed specifically against the native protein of a pathogen. It should be also possible to use such contractions to measure the antibodies directed against the native conformation of a peptidic epitope. All these may have implications for vaccine construction and for the design of immunological tests.

MalE for Production and Purification of Foreign Proteins

General features of MalE hybrid proteins: Random insertion of linkers into malE also generated convenient cloning sites to fuse foreign proteins to the C-terminus or to the N-terminus of the maltose binding protein. Fusion proteins can be recovered from bacterial extracts. As will be discussed, these hybrid proteins display most or all of the properties of each constituent of the fusion.

Firstly, hybrid proteins can be exported into the periplasm and released from this cellular compartment by an osmotic shock treatment (Neu et al., 1965). This is the general case when the foreign protein is a naturally exported protein (e.g. S. aureus nuclease A; Duplay et al., 1987), alkaline phosphatase (Diguan et al., 1988) or soluble CD4 (Clement et al., 1989; Szmelcman et al., 1990). On the contrary, cytoplasmic proteins are much less or even not exported when fused to MalE (e.g. Klenow enzyme or β-galactosidase from E. coli (Duplay et al, 1987; Miki et al., 1987). However we found that subfragments of such proteins can be quite efficiently exported. Membrane bound proteins, when fused to MalE, are often found associated with bacterial membranes (unpublished data). The MalE signal sequence is necessary for export (Duplay et al., 1987), but can be replaced by heterologous signal sequences, of either bacterial (di Guan et al., 1988) or eukaryotic (Clement et al., 1989;

Szmelcman et al, 1990) origin. The possibility of driving foreign proteins into the periplasm is beneficial since this cellular compartment seems better suited than cytoplasm for a correct folding of certain proteins and the establishment of disulfide bridges (Duplay et al., 1987; Miki et al., 1987).

Secondly, the structure of the MalE part is preserved in hybrid proteins since in all cases tested, they can be purified on cross-linked amylose columns, exactly as in the case of wild type MalE. This purification is of particular interest since it is very simple to operate, involves non denaturing steps (elution with 10mM maltose, whereas affinity purification usually necessitates drastic treatments such as low pH) and yields highly purified preparations in one step.

Thirdly, it was found that the biological activity of the protein fused to MalE was preserved when this could be measured (Duplay et al., 1987; Clement et al., 1989; Szmelcman et al., 1990).

Proteolysis of the hybrid proteins occurs during extraction and purification steps rather than during synthesis (unpublished observations): SDS-polyacrylamide gel electrophoresis of material eluted from amylose columns reveals degradation products of discrete sizes ranging from the size of MalE to the expected size of the fusion protein. Proteolysis can be limited to some extent by the addition of chemical inhibitors or by the use of appropriate protease-deficient bacterial strains (Clement et al., 1989; Szmelcman et al., 1990).

Fusions of MalE to CD4, the receptor for HIV: CD4 is an immunoglobulin-like protein which is present at the surface of CD4/helper human lymphocytes. This molecule has been shown to be used by HIV to bind to the surface of lymphocytes through a specific interaction of gp 120 with the first domain of CD4 (four N-terminal domains are localized outside the cell and a transmembrane segment allows its anchoring at the surface of the lymphocyte) (Moore et al., 1990; Ashorn et al., 1990). The soluble part of CD4 has been shown to neutralize HIV in vitro and is currently tested as an antiviral agent.

We made several recombinant molecules between MalE and the soluble region of CD4. In all cases, the signal sequence of MalE was present on the N-terminal side of the fusions. The first two N-terminal domains of CD4 were linked to the N- or to the C-terminus of MalE (MalE-CD4 and CD4-MalE). In more complex constructions, CD4 was fused twice on both sides of MalE (CD4-MalE-CD4, MalE-CD4-CD4 and CD4-CD4-MalE). In vitro recombinations generated chimaeric molecules with one MalE linked to three or four CD4 sequences (CD4-CD4-MalE-CD4-CD4, CD4-CD4-MalE-CD4 and CD4-CD4-MalE-CD4-CD4). The largest proteins we made consisted in two or three malE sequences separated by two CD4 sequences (MalE-CD4-Mal-CD4 and MalE-CD4-MalE-CD4-MalE). A great deal of variability was observed concerning the production and the export of the various proteins. The more complex proteins seemed generally more difficult to express than the short ones.

MER Protein: We studied in detail the expression, purification and properties of MER, an hybrid molecule in which the 177 N-terminal residues of CD4 (i.e. the first two domains) were fused to the C-terminus of MalE (Fig. 2). The protein is efficiently exported into the periplasm. Cells obtained from large cultures of bacteria were first submitted to an osmotic shock which liberates the periplasmic components. The fluid was then concentrated and applied onto a cross-linked amylose column. After extensive wash, the protein was eluted with 10 mM maltose.

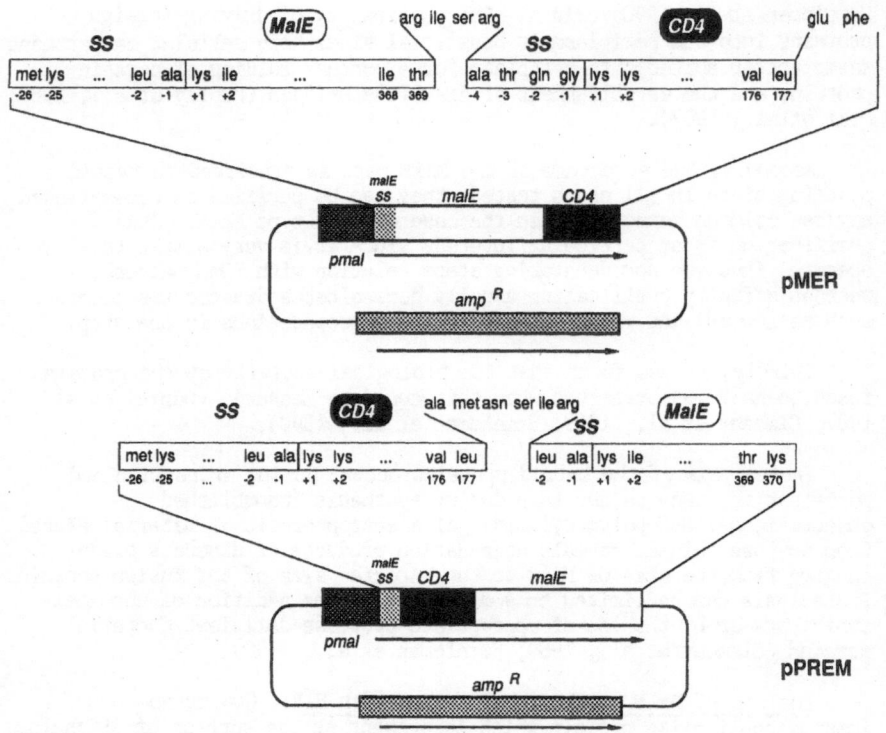

Fig. 2. MalE as an exportation and purification vector.
Two recombinant proteins between MalE and the soluble part
of CD4 are represented. MER: the 177 NH2- terminal
residues of CD4 were fused at the COOH terminal end of the
MalE protein. PREM: The 177 NH2- terminal residues of CD4
were inserted in the region between MalE and the MalE
signal peptide (SS) (see Szmelcman et al., 1990 for
details).

Several grams of pure hybrid protein could be obtained this way starting
from 100 l of bacterial culture (in collaboration with Rhone-Poulenc
Sante).

We could show that the hybrid was recognized by anti-CD4 monoclonal
antibodies exactly as CD4 preparations obtained from other sources did,
bound HIV gp 120 with the same efficiency, and inactivated HIV in vitro
with the same characteristics. This confirmed that MER adopted the
correct conformation in its MalE moiety (binding of amylose and maltose
as an example) as well as in its CD4 part.

The availability of large quantities of MER will allow us to pursue
several studies. MER constitutes a valuable reagent for the analysis of
CD4-gp 120 interaction and potential inhibitors. It can be tested as an
antiviral agent. Since the X-ray structure of MalE has been determined,
obtaining crystals of the hybrid protein should help solving the
structure of the fused CD4 domain.

ROLE OF THE CONTEXT ON THE IMMUNOLOGICAL PROPERTIES OF AN EPITOPE

As shown above, several permissive sites have been defined in LamB and MalE so that the same peptide can not only be expressed at different positions in the vector proteins (intramolecular targeting), but also at different positions within the cell (intramolecular targeting: outer face of the outer membrane, periplasmic face of the outer membrane, periplasm). By using purified proteins, intact cells, or cell extracts, one may study the role of context in the immune responses (humoral and cellular). Such a combined use of vector proteins can be extended to other compartments of the producing bacteria by using other vector proteins. It can also be extended to other Gram⁻ bacteria such as Salmonella and even to mammalian cells where MalE and LamB have recently been expressed (Jean-Marie Clement, unpublished results). This opens, in particular, the possibility to elicit and study cellular immune responses including cytotoxic T responses (CTL) (Aggarwal et al., 1990; Flynn et al., 1990; Moore et al., 1988).

The interpretations can profit from the possibility of determining the structure of the foreign peptide after crystallisation of the hybrid proteins (Rodseth et al., 1990). The more one learns about the vector protein itself in terms of its immunological properties and structure the more one is in a position to relate the properties of the foreign sequences to its context (see sections on MalE and LamB) (Hofnung and Charbit, 1991; Hofnung et al., 1991).

Acknowledgments

This work was supported by grants from World Health Organization (Transdisease Vaccinology Program), Ligue Nationale Francaise contre le Cancer, Fondation pour la Recherche sur le Cancer. D. O'Callaghan was supported by a Wellcome Trust Travelling Fellowship.

References

Aggarwal, A., Kumar, S., Jaffe, R., Hone, D., Gross M., and Sadoff J., 1990, Oral Salmonella: Malaria circumsporozoite recombinants induce specific CD8⁺ cytotoxic T cells, J. Exp. Med., 172:1083.
Ashorn, P., Moss, B., Weinstein, J. N., Chaudhary, V. K., Fitzgerald, D. J., Pastan, I., and Berger, E. A., 1990, Elimination of infectious human immunodeficiency virus from human T-cell cultures by synergistic action of CD4- Pseudomonas exotoxin and reverse transcriptase inhibitors, Proc. Natl. Acad. Sci. USA, 87:8889.
Blondel, A. and Bedouelle, H., 1990, Export and purification of a cytoplasmic dimeric protein by fusion to the maltose-binding protein of Escherichia coli, Eur. J. Biochem., 193:325.
Blondel, A. and Bedouelle, H., 1991, Engineering the quaternary structure of an exported protein with a leucine zipper, Protein Engineering, 4:457.
Charbit, A., Boulain, J. C., Ryter, A., and Hofnung, M., 1986, Probing the topology of a bacterial membrane protein by genetic insertion of a foreign epitope; Expression at the cell surface, EMBO J., 5:3029.
Charbit, A., Gehring, K., Nikaido, H., Ferenci., T. and Hofnung, M., 1988a, Maltose transport and starch binding in phage resistant point mutants of maltoporin functional and topological implications, J. Mol. Biol., 210:487.
Charbit, A., Molla, A., Saurin, W. and Hofnung, M., 1988b, Versatility of a vector for expressing foreign polypeptides at the surface of Gram⁻ bacteria, Gene, 70:181.

Charbit, A., Leclerc, C., Van der Werf, S., Martineau, P., Ronco, J., O'Callaghan, D., and Hofnung M., 1990, Antibody response to foreign epitopes expressed within envelope proteins of Gram⁻ bacteria, in Vaccines 90, eds., F. Brown, R. Chanock, H. Ginsberg and R. Lerner, Cold Spring Harbor Laboratory, New York.

Charbit, A., Ronco, J., Michel, V., Werts, C., and Hofnung, M., 1991, Permissive sites and the topology of an outer-membrane protein with a reporter epitope, J. Bacteriol., 173:262.

Clement, J. -M., Szmelcman, S., Jehanno, M., Martineau, P., Schwartz, O., and Hofnung, M., 1989, Propriétés neutralisantes pour le virus HIV d'une protéine hybride MalE-CD4 exprimée chez Escherichia coli et purifiable en une étape, C.R. Acad. Sci. Paris, 308:401.

Cwirla, S. E., Peters, E. A., Barrett, R. W., and Dower, W. J., 1990, Peptides on phage: a vast library of peptides for identifying ligands, Proc. Natl. Acad. Sci. USA, 87:6378.

Diguan, C., Li, P., Riggs, P. D., and Inouye, H., 1988, Vectors that facilitate expression and purification of foreign peptides in Escherichia coli by fusion to maltose-binding protein, Gene, 67:30.

Duplay, P., Szmelcman, S., Bedouelle, H., and Hofnung, M., 1987, Silent and functional changes in the periplasmic maltose binding protein of Escherichia coli K12.1: Transport of maltose, J. Mol. Biol., 194:663.

Flynn, J. L., Weiss, W. R., Norris, K. A., Seifert, H. S., Kumar, S., and So, M., 1990, Generation of a cytotoxic T-lymphocyte response using a Salmonella antigen-delivery system, Mol. Microbiol., 4:2111.

Hofnung, M., Bedouelle, H., Boulain, J. -C., Clement, J. -M., Charbit, A., Duplay, P., Gehring, K., Martineau, P., Saurin, W., and Szmelcman, S., 1988, Genetic approaches to the study and the use of proteins: random point mutations and random linker insertions, Bull. Inst. Pasteur, 86:95.

Hofnung, M., 1988, Engineered protein fusions and live recombinant bacterial vaccines, in: "Proceedings of the 8th International Biotechnology Symposium", eds., Durand, G., Bobichon, L., Florent, J., Societe Francaise de Microbiologie, Paris, 1988.

Hofnung, M. and Charbit, A., 1991, Expression of antigens as recombinant proteins, in: "Structure of Antigens", ed., M. H. V. Van Regenmortel, The Telford Press, New Jersey.

Hofnung, M., Dougan G., Lanzavecchia, P., and Leclerc, C., 1990, Immumne reponse to proteins with recombinant epitopes, perspectives for vaccines, in: Conferences Philippe Laudat, INSERM, Paris.

Hofnung, M. 1991, Expression of foreign polypeptides at the Escherichia coli cell surface, in: "Methods in Cell Biology", ed., Alan M. Tartakoff, Academic Press.

Kang, A. S., Barbas, C. F., Janda, K. D., Benkovic, S. J. and Lerner, R., 1991, Linkage of recognition and replication functions by assembling combinatorial antibody Fab libraries along phage⁻ surfaces, Proc. Natl. Acad. Sci., USA, 88:4363.

Klauser, T., Pohlner, J., and Meyer, T.F., 1990, Extracellular transport of cholera toxin B subunit Neiseria IgA protease -domain: conformation-dependent outer membrane translocation, EMBO J., 9: 1991.

Leclerc, C., Charbit, A., Molla, A., and Hofnung, M., 1989, Antibody response to a foreign epitope expressed at the surface of recombinant bacteria: importance of the route of immunization, Vaccine, 7:242.

Leclerc, C., Martineau, P., Van der Werf, S., Deriaud, E., Duplay, P., and Hofnung, M., 1990, Induction of virus neutralizing antibodies by bacteria expressing the C3 poliovirus epitope in the periplasm, J. Immunol., 144:3174.

Martineau, P., Szmelcman, S., Spurlino, J. C., Quiocho, F. A., and Hofnung, M., 1990, Genetic approach to the role of tryptophan residues in the activities and fluorescence of a bacterial periplasmic maltose binding protein, J. Mol. Biol., 213:337.

Martineau, P., Charbit, A., Leclerc, C., Werts, C., O'Callaghan, D., and Hofnung, M., 1991, A genetic system to elicit and monitor anti-peptide antibodies without peptide synthesis, Bio/Technology, 9: 170.

McCafferty, J., Griffiths, A. D., Winter, G. and Chriswell, D. J., 1990, Phage antibodies: filamentous phage displaying antibody variable domains, Nature, 348:552.

Miki, T., Yasukoshi, T, Nagatani, H., Furono, N., Orita, T., Yamada, H., Imoto, T., and Horiuche, T., 1987, Construction of a plasmid vector for the regulatable high level expression of eukaryotic genes in Escherichia coli: an application to overproduction of chicken lysozyme, Protein Engineering, 1:327.

Moore, M. W., Carbone, F. R., and Bevan, M. J., 1988, Introduction of soluble protein into the class I pathway of antigen processing and presentation, Cell, 54:777.

Moore, J. P., McKeating, J. A., Weiss, R. A., and Sattentau, Q. J., 1990, Dissociation of gp 120 from HIV-1 virions induced by soluble CD4, Science, 250:1139.

Neu, H. C., and Heppel, L. A., 1965, The release of enzymes of E. coli by osmotic shock during the formation of spheroplasts, J. Biol. Chem., 240:3685.

O'Callaghan, D., Charbit, A., Martineau, P., Leclerc, C., Van der Werf, S., Nauciel, C., and Hofnung, M., 1990, Immunogenicity of foreign peptide epitopes expressed in bacterial envelope proteins, Res. Microbiol., 141:963.

Rodseth, L. E., Martineau, P., Duplay, P., Hofnung, M., and Quiocho, F. A., 1990, Crystallization of genetically engineered active maltose-binding proteins, including an immunogenic viral epitope insertion, J. Mol. Biol., 213:607.

Scott, J. K., and Smith, G. P., 1990, Searching for peptide ligands with an epitope library, Science, 249:386.

Spurlino, J. C., 1988, The 3-D structure of D-maltose binding protein from E. coli., Ph.D. thesis, Rice University, Houston, TX, USA.

Spurlino, J. C., Lu, G. Y., and Quiocho, F. A., 1991, The 2.3-A resolution structure of the maltose-binding or maltodextrin-binding protein, a primary receptor of bacterial active transport and chemotaxis, J. Biol. Chem, 266:5202.

Szmelcman, S., Clement, J. -M., Hehanno, M., Schwartz, O., Montagnier, L., and Hofnung, M., 1990, Export and one-step purification from Escherichia coli of a MalE-CD4 hybrid protein that neutralizes HIV in vitro, Journal of Acquired Immune Deficiency Syndrome, 3:859.

Participants of the NATO Advanced Studies Institute "Targeting of Drugs: The Challenge of Peptides and Proteins" held at Cape Sounion Beach, Greece, during 24 June–5 July, 1991. The Organizing Committee included Gregory Gregoriadis (ASI Director and Chairman), Alexander T. Florence (ASI Co-Director), George Poste (ASI Co-Director), Theo J.C. van Berkel and Brenda McCormack (ASI Co-Ordinator).

CONTRIBUTORS

A.C. Allison, Institute of Immunology and Biological Sciences, Syntex Research, Palo Alto, CA 94304, USA

D.R. Bard, Strangeways Research Laboratory, Worts Causeway, Cambridge CB1 4RN, UK

L. Benatti, Biotechnological Research, Farmitalia Carlo Erba, 24 Via E. Bezzi, 20146 Milano, Italy

N.E. Byars, Institute of Immunology and Biological Sciences, Syntex Research, Palo Alto, CA 94304, USA

A. Ceriotti, Istituto Biosintesi Vegetali del CNR, Via Bassini 15, 20133 Milano, Italy

A. Charbit, Unite de Programmation Moléculaire and Toxicologie Génétique (CNRS URA 1444), Institut Pasteur, 25 rue du Dr. Roux, 75015 Paris, France

J.-M. Clément, Unité de Programmation Moleculaire and Toxicologie Genetique (CNRS URA 1444), Institut Pasteur, 25 rue du Dr. Roux, 75015 Paris, France

D.J.A. Crommelin, Department of Pharmaceutics, University of Utrecht, P.O. Box 80.082, The Netherlands

W.M.C. Eling, Department of Medical Parasitology, University of Nijmegen, Nijmegen, The Netherlands

A. Engert, Medizinische Universitatsklinik I der Universitat zu Koln, 5000 Koln 41, Germany

A. Gabizon, Sharett Institute of Oncology, Hadassah University Hospital, Jerusalem, Israel

G. Gregoriadis, Centre for Drug Delivery Research, School of Pharmacy, University of London, 29-39 Brunswick Square, London WC1N 1AX, UK

M. Hofnung, Unité de Programmation Moléculaire and Toxicologie Génétique (CNRS URA 1444), Institut Pasteur, 25 rue du Dr. Roux, 75015 Paris, France

L. Huang, Department of Pharmacology, University of Pittsburgh, School of Medicine, Pittsburgh, PA 15261, USA

S.K. Huang, Cancer Research Institute, University of California, San Francisco, CA, USA

C.G. Knight, Strangeways Research Laboratory, Worts Causeway, Cambridge CB1 4RN, UK

D.A. Lappi, Department of Cellular Growth Biology, The Whittier Institute for Diabetes and Endocrinology, La Jolla, CA, USA

C. Leclerc, Biologie des Régulations Immunitaires, Institut Pasteur, 25 rue du Dr. Roux, 75015 Paris, France

P. Martineau, Unite de Programmation Moleculaire and Toxicologie Genetique (CNRS URA 1444), Institut Pasteur, 25 rue du Dr. Roux, 75015 Paris, France

T.S. Maughan, Velindre Hospital, Whitchurch, Cardiff, CF4 7XL, UK

S. Muir, Unité de Programmation Moléculaire and Toxicologie Génétique (CNRS URA 1444), Institut Pasteur, 25 rue du Dr. Roux, 75015 Paris, France

D. O'Callaghan, Unité de Programmation Moléculaire and Toxicologie Génétique (CNRS URA 1444), Institut Pasteur, 25 rue du Dr. Roux, 75015 Paris, France

D.P. Page-Thomas, Strangeways Research Laboratory, Worts Causeway, Cambridge CB1 4RN, UK

D. Papahadjopoulos, Cancer Research Institute, University of California, San Francisco, CA, USA

P.A.M. Peeters, Pharma bio-Research Consultancy BV, Zuidlaren, The Netherlands

O. Popescu, Unité de Programmation Moléculaire and Toxicologie Génétique (CNRS URA 1444), Institut Pasteur, 25 rue du Dr. Roux, 75015 Paris, France

E. Rozengurt, Imperial Cancer Research Fund, P.O. Box 123, Lincoln's Inn Fields, London WC2A 3PX, UK

M. Saffran, Department of Biochemistry and Molecular Biology, Medical College of Ohio, Toledo, Ohio 43699, USA

T. Sethi, Imperial Cancer Research Fund, P.O. Box 123, Lincoln's Inn Fields, London WC2A 3PX, UK

M.R. Soria, Department of Biotechnology, San Raffaele Research Institute, Via Olgettina 60, 20133 Milano, Italy

S. Szmelcman, Unité de Programmation Moléculaire and Toxicologie Génétique (CNRS URA 1444), Institut Pasteur, 25 rue du Dr. Roux, 75015 Paris, France

P. Thorpe, Cancer Immunobiology Center, The University of Texas, Southwestern Medical Center, Dallas, Texas 75235-8576, USA

A. Vitale, Istituto Biosintesi Vegetali del CNR, Via Bassini 15, 20133 Milano, Italy

E.P. Wraight, Department of Nuclear Medicine, Addenbrooke's Hospital, Hills Road, Cambridge, CB2 2QQ, UK

F. Zhou, Department of Pharmacology, University of Pittsburgh, School of Medicine, Pittsburgh, PA 15261, USA

INDEX

α-melanocyte stimulating hormone,
    see also MSH, 1, 2
  activity of, 2
  amino-acid sequence, 1
  analogues of, 2
  conjugates of, 2
  derivatives,
    diethylenetriamine penta-
    acetic acid,2
  in drug targeting, 1-
  molecular weight, 2
  targeting to melanomas, 2
Angiotensin, 99
Antagonists, 101
  antagonist A, 102
    mechanism of action, 102
  antagonist D, 102
    mechanism of action, 102
  antagonist G, 102
    mechanism of action, 102
  for neuropeptides, 101, 102
  substance P, 102
Antigen presenting cells, 74
  types, 74
Asialofetuin, 47
  half-life, 47
  in liposomes, 47
  receptor for, 47

Bacteria, 113
  as reagents, 113
Bacterial, 110
  envelope proteins, 110
    LamB protein, 110
Bombesin, 98, 100
  action, 98
  as mitogen, 100
  nuclear oncogenes, effect
    on, 100
  receptors, binding to, 99
    intracellular signals, 99
  tyrosine phosphorylation,
    induction of, 99
Bradykinin, 98, 100

Cell-mediated immunity, 63

Cell-mediated immunity (cont'd)
  liposome-induced, 63
Chloroquine, 33
  in liposomes, 33
Cholecystokinin, 100, 101
  calcium, effect on, 100
  tumours, secretion by, 101

Dehydration-rehydration
    vesicles, 60
  drugs entrapped in, 60
    entrapment yield, 60
    technology, 60
Dianthin-30, 23
  intracellular traffic, 23
  propeptides, 23
Dianthus caryophyllus, 19
Diethylenetriamine penta-
    acetic acid, 2
  MSH conjugates, 2
    clinical trials, 4-6
    hormonal activity, 3, 4
Dioleoylphosphatic acid, 46
  in immunoliposomes, 46
Diphtheria toxoid, 62
Doxorubicin, 53
  in liposomes, 53

E. coli, 112
  recombinant, 112
  vaccines, 112
Endothelin, 98, 99
Envelope protein, 110, 113
  LamB protein, 110, 12
    functions, 110
    location, 110
    permissive sites, 110, 111
  MalE protein, 110, 113
    functions, 110
    permissive sites, 110, 111
    purification, 113

Fibroblast growth factor, 24

Galanin, 100
  calcium, effect on, 100

125

Ganglioside GM$_1$, 47
  in liposomes, 47
Genetic engineering, 109
  recombinant proteins, 109
Gold, 53
  colloidal, 53
  in liposomes, 53
Growth factors, 97, 98, 100
  autocrine, 97
  interactions between, 97
  neuropeptides, 98
  nuclear oncogens, effect on,
    10
  paracrine, 97
Herpes simplex virus antigen, 74
  in Syntex adjuvant formul-
    ation, 74
High density lipoproteins, 61
  liposomes, effect on, 61
Hodgkin tumors, 12
  immunotoxins, treatment with,
    12-14
  in nude mice, 12
Humoral immunity, 63
  liposome-induced, 63

IgA protease, 109
  domains, 110
  IgA antibodies, effect on, 110
Immunoliposomes, 31-34, 38, 39,
    45, 46
  antibody preparation, 33
  cell targets for, 32
  chloroquine-laden, 33
    preparation, 34
  destabilization, 46
  dragging concept, 32
    therapeutic potential, 33
  Fab'-coated, 38
  in imaging, 32
  in vivo studies, 34-39
  long-circulating, 46
  peptides, delivery of, 45
  proteins, delivery of, 45
  stabilizers, 46
    acylated antibody, 46
  target-sensitive, 45, 46
  targeting in vivo, 37-42
    to circulating cells, 31
    to erythrocytes, 37-39
    to reticulocytes, 39,41
Immunological adjuvants, 71
  alanyl-MDP, 71
  threonyl-MDP, 71
  cytokine induction, 59, 69
  Freund's complete adjuvant, 69
  Freund's incomplete adjuvant, 69
  isotypes formed, 69
  liposomes, 59-66

mechanism of action, 59
  muramyl dipeptide analogues,
    69
  toxicity, 60
Immunotoxins, 9, 11, 12, 14
  antitumor effects, 14,15
  crossreactivity, 11, 12
  in Hodgkin's tumors, 12
  potency of, 9, 11
  purification, 11
  ricin A-chain, 9
  specificity, 11
  therapy with, 9
  toxicity, 11
Insulin, 82-84, 89-93
  arterial glucose, effect on,
    91
  in diabetic dogs, 93
  fate after injection, 90, 91
    in diabetics, 90
    side effects, 90, 91
  glucose levels, 93, 94
  in liposomes, 83, 84
  liver, effect on, 91
  in microspheres, 84
  muscle cells, effect on, 91
  oral administration, 82
    absorption, 82
    blood levels, 82, 83
    diabetic rats, 82
    in gelatin capsules, 92
    glucose levels, 82, 83
    multiple doses, 92
    protection, 82, 83, 84
Interdigitating cells, 74
Interleukin-2, 63
  in liposomes, 63

LamB protein, 110, 112
  functions, 110
  genetic coupling to, 112
    applications, 112, 113
    of viral epitopes, 112
  location, 110
  permissive sites, 110, 111
Langerhans cells, 74
Ligand toxins, 19
  design of, 19
Lipid polymorphism, 45, 46
  and liposomal stability, 45,
    46
  phosphatidylethanolamine, 45,
    46
Lipopolysaccharide, 70
  as adjuvant, 70
  toxicity, 70
    arthritis, 70
    pyrogenicity, 70
    uveitis, 70

Liposomal, 47, 59
  asialofetuin, 47
  ricin, 47
  vaccines, 59-66
Liposomal adjuvanticity, 63
  amplification, 63
    with co-adjuvants, 63
    with interleukin-2, 63
Liposome-entrapped, 34, 45, 53,
    56, 62, 63, 83
  antigens, 62
  chloroquine, 34
  diphtheria toxoid, 62
  doxorubicin, 53, 56
    tumour distribution, 53-55
    therapeutic efficacy, 57
Liposome-entrapped (cont'd)
    toxicity studies, 56
  gold, 53
    biodistribution studies,
      53-55
    morphological studies, 53
    therapeutic efficacy, 57
    toxicity studies, 56
  insulin, 83, 84
  interleukin-2, 63
  peptides, 45
  proteins, 45
Liposomes, 37, 45, 47, 48, 51
    52, 59, 60-62, 64, 65
  behaviour in vivo, 61
    intragastric route, 62
  cells, uptake by, 62
    lysosomotropic, 62
  dehydration-rehydration
      vesicles, 60
  as drug carriers, 51
    in tumour-bearing mice, 51
  Fab'-coated, 37
  half-lives, 61
  and high density lipoproteins,
      61
  immunoadjuvant action, 64
    bilayer fluidity, role of,
      64-65
    control, 64
    structural characteristics,
      role of, 64
  as immunological adjuvants, 59,
      62, 65, 62-66
    antibody production, 63
    antibody subclasses, 63
    B-cell epitotes, 65
    cell-mediated immunity, 63
    for hepatitis B surface
      antigen, 65
    humoral immunity, 62
    mechanism, 62
    for peptides, 65
    pre-S$_2$ peptide, 65

S peptide, 65
  T-cell epitopes, 65
microfluidization of, 61
peptides, delivery of, 45
pH-sensitive, 47, 48
  antigen presenting cells,
    action in, 48
proteins, delivery of, 45
stability, 61
Stealth^R, 51, 52
  design, 51
  mechanisms of function, 52
  negative charge, the role
    of, 52
sterically stabilized, 51
  criteria for, 51
technology, 60
MalE protein, 113-115
  application, 114
    peptide delivery, 114
    protein production, 114
    protein purification, 114
  functions, 113
  fusion to CD4, 115, 116
  hybrid proteins, 114
  location, 113
Microfluidization, 61
  of liposomes, 61
Monoclonal antibodies, 1, 9
  against CD determinants, 9
    crossreactivity, 9
  distribution in tumours, 1
Muramyl dipeptide(MDP), 70, 71
  as adjuvant, 70
  analogues, 71
    alanyl-MDP, 71
    aminobutyryl-MDP, 71
    desmethylalanyl-MDP, 71
    threonyl, 71
  structure, 70
  synthetic analogues, 70
  toxicity, 70
    arthritis, 70
    pyrogenicity, 70
    uveitis, 70

Neuropeptides, 97-104
  angiotensin, 99
  antagonists, 101, 103
    broad spectrum, 103
    effects of, 103
    non-antigenic, 101
    structure, 102
    testing of, 101
  bombesin, 98
  bradykinin, 98, 100
  in calcium mobilization, 100
  cholecystokinin, 100
  endothelin, 98, 99
  galanin, 100

Neuropeptides (cont'd)
  neurotensin, 100
  serothelin, 99
  substance K, 99
  substance P, 99
  vasoactive intestinal contrac-
    tor, 98
  vasoactive intestinal peptide,
    98
  vasopressin, 98, 100
Neurotensin, 100, 101
  calcium, effect on, 100
  tumours, secretion by, 101

Oil-in-water emulsions, 72
  toxicity, 72

Pegylated lipids, 51
  in liposomes, 51
Peptide hormones, 103
Peptides, 46, 81, 82, 85
  colon, delivery to the, 85,
    86
    strategies, 85, 86
  colon, protection in, 85
  half-lives, 46
  oral administration, 81, 82
    absortion, 82
    blood levels, 82
    diabetic rats, 82
    glucose levels, 82, 83
    of insulin, 82
  small intestine, protection
    in, 81-83
Permissive sites, 109
  detection, 109
Phosphatidylethanolamine, 45
  unsaturated, 45
    polymorphism of, 45
Phytolacca americana, 19
Phytolacca dodecandra, 19
Plant toxin gene, 24
  cloning of, 24
  expression of, 24
Pluronic copolymers, 71-77
  copolymer L-121, 71
    structure, 71
  polyoxyethylene polypropylene
    copolymer, 72
Polyethyleneglycol, 51
  in liposomes, 51
Polypeptides, 109
  purification, 109
  targeting of, 109
Protein domains, 110
  insertion of, 110
    applications, 112, 113
    methodology, 110

Proteins, 46, 109, 110
  bacterial envelope, 110
  domains, 110
  half-lives, 46
  outer membrane, 110
  periplasmic maltose binding,
    110
  as vectors, 109

Recombinant proteins, 109
  applications, 109, 112, 113
  construction, 109
Ribosome-inactivating proteins,
  19
  amino acid sequences, 19
  cloning of, 19-22
  dianthin-30, 19
  from Saponaria officinalis, 19
  saporin-6, 19
  single chain, 19
  sources, 19
Ricin-A-chain, 9, 11
Ricin-A-chain (cont'd)
  deglycosylated, 9, 11
  immunotoxins, 9

Salmonella, 112
  recombinant, 112
    vaccines, 112
Saporin-6, 19, 21, 23, 24, 25
  cloning of, 19
  expression in E. coli, 24
  glycosylation, 21
  immunoconjugates, 25
    cytotoxicity, 25
    receptor-targeted, 25
  intracellular traffic, 23
  propeptides, 23
    glycosylation site, 23
Saporin-based, 24
  immunoconjugates, 24
    with fibroblast growth fac-
      tor, 24, 25
Saponaria officinalis, 19
Serotonin, 99
Shigela, 112
  recombinant, 112
    vaccines, 112
Squalane, 72
  in adjuvant formulations, 72
  oil-in-water emulsions, 72
  safety, 72
Squalene, 72
  in adjuvant formulations, 72
  oil-in-water emulsions, 72
  safety, 72
Substance K, 99
Substance P, 99

Syntex adjuvant formulation, 72
        74, 76
    as adjuvant, 72
    antigens in, 74
    composition, 72
    cytokine production, 76
    helper cell responses, 76
    herpes simplex antigen in, 74
    isotype production, 76
    mechanism of action, 72, 74
    MHC-class II restricted cyto-
        toxicity, 75
    structure, 73
    in vaccines, 76
        against HIV, 76
        against influenza, 77
        against simian AIDS virus,
            76

Tumours, 97
    growth, 97

Vaccines, 59, 69, 112
    hybrid proteins, 112
        immunogenicity, 112
    liposome-based, 59-66
    live bacterial, 112
        E. coli, 112
        immunogenicity, 112
        Salmonella, 112
        Shigela, 112
    multivalent, 112
    subunit, 69
Vasoactive intestinal contractor,
    98
Vasoactive intestinal peptide,
    98
Vasopressin, 98-101
    tumours, secretion by, 101
    tyrosine phosphorylation,
        induction of, 99

Water-in-oil emulsions, 72
    toxicity, 72